The Akasha Paradigm in Science

(R)EVOLUTION AT THE CUTTING EDGE

Akashic Science books by Ervin Laszlo

l'Ipotesi del Campo Psi (in Italian)
(Lubrina, 1988)

The Creative Cosmos
(Floris Books, 1993)

The Interconnected Universe
(World Scientific, 1995)

The Whispering Pond
(Element Books, 1996)

The Connectivity Hypothesis
(State University of New York Press, 2003)

Science and the Akashic Field
(Inner Traditions, 2004, 2007)

Quantum Shift in the Global Brain
(Inner Traditions, 2008)

Cosmos (with Jude Currivan)
(Hay House, 2008)

The Akashic Experience
(Inner Traditions, 2009)

The Birth of the Akasha Paradigm
(ibook, 2012, akashaparadigm.com)

The Definitive Theory by the Bestselling Author of
Science and the Akashic Field

The Akasha Paradigm in Science

(R)EVOLUTION AT THE CUTTING EDGE

Ervin Laszlo
with contributions by

Peter Jakubowski
Paul A. La Violette
László Gazdag
James D. Bourassa
and
David W. Thomson III

Waterside Press

Published by Waterside Publications
ISBN: 978-1-937504-18-2
WPP134P

Manufactured in the United States of America or in Great Britain, when purchased outside of North or South America

Produced and distributed for
Waterside Publications by
Worthy Shorts Publisher Services BackOffice
A CustomWorthy edition

For further information contact
info@worthyshorts.com

For Carita Marjorie

Without whose unfailing patience and constant love and support I could not have had the endurance, the inspiration and the concentration to live and breathe Akasha while this ancient insight found its way into concepts that can articulate the new paradigm struggling to be born in both science and society.

Acknowledgements

I am deeply grateful to Edgar Mitchell, Deepak Chopra, David Loye, Stanley Krippner, and Kingsley Dennis, long-standing friends and colleagues for their comments and suggestions, invaluable resources for achieving what I now consider the definitive formulation of Akashic theory.

I wish to thank my literary agent, friend and colleague and now also publisher Bill Gladstone for his sincere and unwavering belief in the message conveyed in this book;

my dedicated executive assistant Györgyi Szabo, for researching the reference materials for this study;

and the physicists who with their brilliant hypotheses have helped lay the foundations of the Akasha paradigm in physics:

Peter Jakubowski, László Gazdag, Paul A. La Violette, James D. Bourassa, and William Thomson III.

Contents

PREFACE (R)Evolution in Science ix

Part I
THE BREAKDOWN OF THE CURRENT PARADIGM 1
1. Anomalies in the Quantum World 3
2. Anomalies in the Living World 11
3. Anomalies in the Universe 25

Part II
INTRODUCING THE AKASHA PARADIGM 37
4. Fields 39
5. The Akasha 47

Part III
THE AKASHIC VIEW OF THE WORLD 53
6. Cosmos 55
7. Cognition 61
8. Freedom 67
9. The Good 71
10 Consciousness 75

Annexes
THE AKASHIC (R)EVOLUTION IN PHYSICS 83
Four Hypotheses
I. The Universal Quantum Field: Foundations of the Unified Description of Basic Physical Quantities
Peter Jakubowski 85

II. The Superfluid Ether as the Basis for Unifying the Fields and Forces of Nature
László Gazdag 103

III. The Transmuting Ether: Introduction to Subquantum Kinetics
Paul A. LaViolette 117

IV. The Aether Model: A New Foundation for Physics
David W. Thomson III and James D. Bourassa 143

SUMMING UP 151

COMMENTS 159
Edgar Mitchell, David Loye, Kingsley Dennis, Stanley Krippner, Deepak Chopra

Preface

(R)EVOLUTION IN SCIENCE

Einstein said, "We are seeking the simplest possible scheme of thought that can tie together the observed facts." This phrase encapsulates the quintessence of the project we know as science. Science is not technology, not even discovery: it is understanding. When our understanding of the world matches the nature of the world we will discover more and more of the world, and will have more and more ability to cope with it. Understanding is basic.

Genuine science seeks the scheme that could convey comprehensive, consistent, and optimally simple understanding. That scheme is not established once and for all; it needs to be periodically updated. The observed facts grow with time and become more diverse. Tying them together in an optimally simple yet comprehensive scheme calls for revising and occasionally re–inventing that scheme. In recent years the repertory of observed facts has grown and has become highly diverse. We need a new scheme: a new paradigm. This book seeks to contribute to the advance of science by investigating the nature of that paradigm.

Science evolves through alternating phases of what Thomas Kuhn called normal science, and the nonlinear processes constituted by shifts in the dominant paradigm. Paradigm shifts are radical steps: they are a revolutionary form of evolution, a (r)evolution. Normal science treads water: it is only marginally innovative. It ties together the observed facts within an established and consensually validated scheme, and if it encounters observations that do not fit that scheme, it extends and adjusts that scheme. This, however, is not always possible. If the attempt is not relinquished, the dominant scheme becomes unmanageably complex and opaque, as Ptolemaic astronomy did through the constant addition of epicycles to its basic cycles to account for the "anomalous" movement of the planets. When in the growth of science that critical point is reached, it is time to take the radical (r)evolutionary step. The dominant scheme must be replaced. A new paradigm must be found to ground the theories and interpret the observations that support them.

In the natural sciences this point has now been reached. A number of unexpected, and for the current paradigm critically anomalous observations have come to light.

The series of critically anomalous observations can be traced to an experimental finding in 1982. A paper by French physicists Alain Aspect and collaborators reported on an experiment carried out under rigorously controlled conditions. This experiment demonstrated that particles that are split, and of which the split halves are projected a finite distance from each other, remain connected despite the space that separates them. Moreover their connection seems to be instant. This contradicts a basic tenet of relativity: according to Einstein's theory the speed of light is the highest speed at which any thing or signal can propagate in the universe.

Aspect's experiment was repeated, and always produced the same result. The science community was baffled, but finally dismissed the phenomenon as not having deeper significance: the "entanglement" of the split particles, physicists said, is strange, but it does not convey information or "do" anything. But this, too, was placed in question in subsequent experiments. It turned out that the quantum state of particles, and even of whole atoms, can be instantly projected across any finite distance. This came to be known as "teleportation." Then instant quantum–resonance–based interactions were discovered also in living systems, and even in the universe at large.

A related anomalous fact came to light in regard to the level and form of coherence found in complex of systems. The observed coherence suggests "nonlocal" interaction between the parts or elements of the systems: interaction that transcends the recognized bounds of space and time. This kind of interaction surfaced not only in the quantum domain but, surprisingly, also at macroscopic dimensions.

Yet another finding inexplicable by the current paradigm is that organic molecules are produced in stars. The received wisdom is that the universe is a physical system in which life is, if not an anomalous, at least a rare and very likly accidental phenomenon. After all, living system can evolve only under conditions that are extremely rare in space and time. However, it turned out that the basic substances on which life is based are produced already in the physical–chemical evolution of stars. These organic molecules are ejected into surrounding space and coat asteroids and clumps of interstellar matter, including those that subsequently condense into stars and planets. It appears that the universe is distinctly tuned for life: its basic physical processes produce the very substances required for the evolution of living systems.

Observations of this kind are not amenable to being tied together by patching up the dominant scheme: they challenge not only the maximum velocity of effect–propagation in space, but our understanding of the most basic processes in nature. They place in question the current paradigm: the fundamental scheme by which scientists tie together the observed facts. This was the case at the turn of the 20th century, with the shift from the Newtonian to the relativity paradigm, and again in the 1920s with the advent of quantum theory. More limited (r)evolutions have unfolded in specific domains since then, among them in psychology with the emergence of transpersonal theories and in cosmology as shown by the advent of non–Big Bang multicyclic models. The next (r)evolution promises to be encompassing again: shifting from the still dominant local–state paradigm to a paradigm based on nonlocal interaction and systemic coherence: a paradigm of embracing wholeness in nature, with important implications for human life and aspiration.

This book introduces the basic contours of the indicated nonlocal–state "Akasha" paradigm and explores its implications for our view of the world and our behavior in the world.

—The observations that trigger the search for a new paradigm are surveyed in Part I.

—The basic assumptions of the new "Akasha" paradigm are the subject of Part II.
—Part III explores the power of the Akasha paradigm to ground a coherent integral view of the world.
—The mathematical hypotheses of the Annex explore the foundations of the Akasha paradigm in physics.
—A concise yet comprehensive Summing Up followed by five illuminating Comments concludes the book.

PART I

THE BREAKDOWN OF THE CURRENT PARADIGM

Credible theories of the world must be based on observations of the world. Theory and observation are inherently intertwined. There is no "immaculate perception"—everything we see and hear is colored by what we believe about the things and events we see and hear.

But not all we see and hear serves as a basis for a credible theory; our observations must hang together in a comprehensible manner. Theories can tie the observations together in most cases, but not in all. Some observations remain recalcitrant: they appear to refuse, or at least to resist, integration within a theory—they are anomalous in regard to its basic tenets. If the anomalies persist, there is a need either to review the validity of the observations—they could be mistaken—or revise the theory by which they are interpreted. If the observations are checked and repeated and still remain anomalous, innovation in the field of theory is called for.

Observations that are both widespread and anomalous for the basic assumptions of the established theories are particularly vexing. Adjustments in regard to particular findings are often possible, but when they accumulate they render the theories that account for them unmanageably complex. If they challenge their basic assumptions they raise the possibility that the paradigm that grounds those theories has become obsolete. This calls for a radical step: abandoning the current paradigm and finding and then shifting to a new one.

Observations of a surprising level and form of coherence in nature pose a challenge of this fundamental kind. They are widespread. The coherence that appears is not coherence of the ordinary or garden variety: it indicates a non–classical, instant or quasi–instant connection among the parts of the coherent systems. The parts are correlated in such a way that what happens to one also happens to the others, and so it happens to the system as a whole. All the parts respond to the "rest of the world" as a whole, maintain themselves as a whole, and change and evolve as a whole.

The implications of these findings are profound. It appears that particles and atoms, and the macro–scale entities constituted as ensembles of particles and atoms, are not just local. The parts of these entities are not limited to the place where they are found; they encompass the entire system of which they are a part. And the systems themselves are nonlocal: they are not just here and now, but are present in some sense in all systems in space and time. In the final count all things in the world are essentially "nonlocal."

Nonlocality was believed to be limited to the microscale world of the quantum, and even there only to the pristine state of the quanta. But it appears to be widespread. The chapters that follow demonstrate that nonlocality also occurs at the mesoscale of life, and even at the macroscale of the universe. A paradigm that conceives of interactions as local cannot account for this finding; it needs to be reconsidered and replaced.

Chapter One

Anomalies in the Quantum World

The quantum world is intrinsically interconnected, with nonlocal connection between quanta. At the fundamental level, the world is clearly nonlocal.

As we shall see, the phenomenon of nonlocality is fundamental for our understanding of the world not only on the supersmall scale, but at macroscales as well. This is not a multilayer world with different levels obeying their own laws and following their own dynamics; it is an integral, whole–system world where basic processes occur and recur at all scales of size and complexity. The way the fundamental elements of the world are interconnected, and how they interact tell us much about the way this world is built, and how it functions.

The world of the quantum

Quanta, the smallest known entities of the physical world, do not behave like larger–scale objects. Until an instrument or an act of observation registers them, they have neither a unique location nor a unique state. And they are nonlocally connected throughout space and time.

The quantum state is defined by the wavefunction that encodes the superposition of all the potential states a given quantum can occupy. The superposed state of the quantum is the pristine state, in the absence of all interaction. The duration of the pristine state may vary. It can be just the millisecond that it takes a pion to decay into two photons, or it can be ten thousand years in the decay of a uranium atom. Whatever its length, it is the state of superposition known as one tick of a quantum clock, or *q–tick*. According to the Copenhagen interpretation of quantum theory, reality (or at least space and time) does not exist during a q–tick, only at the end of it, when the wavefunction has collapsed and the quantum has transited from the superposed indeterminate to the classical determinate state.

It is not clear what would bring about the collapse of the wavefunction. Eugene Wigner speculated that it is due to the act of observation: the consciousness of the observer interacts with the particle. Yet also the instrument through which the observation is made could impart the crucial impetus, in which case the transition occurs whether or not an observer is present. Heisenberg leaned now to one view and now to the other.

Until the quantum is measured (or possibly just observed), it has the properties of both waves and corpuscles. But, as Niels Bohr's theory of wave–particle complementarity, and Werner Heisenberg's principle of indeterminacy tells us, these properties do not and cannot occur at the same time. The wave–state of the quantum excludes the particle–state, and vice versa. This means that all properties of the quantum cannot be measured at the same time. When one property is measured, another property becomes blurred, or its value goes to infinity.

The nonlocality experiments

Nonlocality in the world of the quantum surfaces in a series of experiments, each more astonishing than the other.

This series of experiments started with the by now "classical" explorations of Thomas Young in the early nineteenth century. Young made coherent light pass through an intervening surface with two slits. He placed a screen behind this surface to receive the light that penetrated the slits. He found that a wave–interference pattern had appeared on the screen. This would suggest that photons, as waves, had passed through both slits. But how could this happen when the light source is so weak that only one photon is emitted at a time? A single packet of light energy should behave as a corpuscular entity; it should be able to pass through only one of the slits. Yet even in this case an interference pattern builds up on the screen. Can even a single photon behave as a wave?

John Wheeler carried out an entire series of more precise and sophisticated experiments in the 1970s and '80s.[1] In his experiments as well, photons are emitted one at a time and are made to travel from the emitting gun to a detector. The latter clicks when a photon strikes it. A half-silvered mirror is inserted along the photon's path; this splits the beam, giving rise to the probability that one in every two photons passes through the mirror and one in every two is deflected by it. To confirm this probability, photon counters that click when hit by a photon are placed both behind this mirror, and at right angles to it. The expectation is that on the average one in two photons will travel by one route and the other by the second route. This is confirmed by the results: the two counters register a roughly equal number of clicks—and hence of photons. When a second mirror is inserted in the path of the photons that were undeflected by the first, one would still expect to hear an equal number of clicks at the two counters: the individually emitted photons would merely have exchanged destinations. But this expectation is not borne out by the experiment. Only one of the two counters clicks, never the other. All the photons arrive at one and the same destination.

It appears that the singly emitted and therefore supposedly corpuscular photons interfere with one another as waves. Above one of the mirrors their interference is destructive—the phase difference between the photons is one hundred eighty degrees—so that the photon waves cancel each other. Below the other mirror the interference is constructive: the wave–phase of the photons is the same and as a consequence they reinforce one another.

Photons that interfere with each other when emitted moments ago in the laboratory also interfere with each other when emitted in nature at considerable intervals of time. The "cosmological" version of Wheeler's experiment bears witness to this. In

1 Wheeler, John A., 1984. "Bits, quanta, meaning," *Problems of Theoretical Physics*, A Giovanni, F. Mancini, and M. Marinaro (eds.) Salerno: University of Salerno Press.

this experiment the photons are emitted not by an artificial light source, but by a distant star. In one experiment the photons of the light beam emitted by the double quasar known as 0957 + 516*A,B* were tested. This distant quasi–stellar object is believed to be one star rather than two, the double image being due to the deflection of its light by an intervening galaxy situated about one fourth of the distance from Earth. (The presence of a galaxy, as of all mass, is believed to curve the space–time matrix, and so curve the path of the light beams that propagate through it.) The deflection due to this "gravitational lens" action is large enough to bring together two light rays emitted billions of years ago. Because of the additional distance traveled by the ray deflected by the intervening galaxy, it has been on the way fifty thousand years longer than the ray that traveled the direct route. But, although originating billions of years ago and arriving with an interval of fifty thousand years, the two light rays interfere with each other just as if the photons that constitute them would have been emitted seconds apart in the laboratory. It appears that whether light particles are emitted at intervals of a few seconds in the laboratory, or at intervals of thousands of years in the universe, those that originated from the same source create wave–interference patterns with each other.

The interference of photons and other quanta is extremely fragile: any coupling with another system destroys it. Still more recent experiments came up with a still more astonishing finding. It turned out that when any part of the experimental apparatus is coupled with the source of the photons, the fringes that indicate interference among them vanish. The photons behave as classical particles.

Further experiments have been devised to determine through which of the slits a given photon is passing. Here a so–called "which–path detector" is coupled to the emitting source. The result is that as soon as the apparatus is in place, the interference fringes weaken and ultimately vanish. The process can be calibrated: the higher the power of the which–path detector, the more of the fringes disappear.

An experiment conducted at the Weizmann Institute in Israel made use of a device less than one micrometer in size. The instrument creates a stream of electrons flowing across a barrier on one of two paths. The paths focus the electron streams and enable the investigators to measure the level of interference between the streams. With the detector turned on for both paths, the interference fringes disappear as expected. And the higher the detector is tuned for sensitivity, the smaller the interference patterns that appear. A surprising factor enters into play: the coupling of the measuring apparatus to the light source.

An even more surprising factor appears as well. In some experiments the interference fringes disappear as soon as the detector apparatus is readied—and even when the apparatus is not turned on. In Leonard Mandel's optical–interference experiments two beams of laser light are generated and allowed to interfere.[2] When a detector is present that enables the path of the light to be determined, the fringes disappear. But they disappear regardless of whether or not the determination is actually carried out. It appears that the very possibility of "which–path–detection" collapses the photons' wavefunction: it destroys their superposed state.

This finding was confirmed in experiments carried out at the University of Konstanz by Dürr and collaborators in 1998.[3] In these experiments the puzzling interference fringes were produced by the diffraction of a beam of cold atoms by standing waves of light. When there is no attempt to detect which path the atoms are taking, the interferometer displays fringes of high contrast. However, when information is encoded within the atoms as to the path they take, the fringes vanish. Yet the instrument itself cannot be the cause of the collapse—it does not deliver a sufficient "momentum kick" since the back action path of the detector is four orders of magnitude smaller than the separation of the interference fringes. In any case, for the interference pattern to disappear, the labeling of the paths does not need to be actually carried out: it is enough that the atoms are labeled so that the path they take *can* be determined. It appears that the measuring apparatus is "entangled" with the object that is measured.

2 Mandel, Leonard, 1991. Physical Review Letters 67:3, 318–321

3 Dürr, S., T Nonn and G. Rempe, 1998. "Origin of quantum–mechanical complementarity probed by a 'which–way' experiment in an atom interferometer" *Nature*, vol. 395 (3 September)

The EPR experiment

Nonlocal connection among quanta that originated in the same quantum state was demonstrated by the famous EPR (Einstein–Podolski–Rosen) thought–experiment. Put forward by Albert Einstein with Boris Podolski and Nathan Rosen in 1935, this was a "thought" experiment because at the time it could not be empirically tested.

Einstein and colleagues suggested that we take a proton or another particle in a so–called singlet state where their spins cancel out each other and yield a total spin of zero. We then allow the particles to separate and travel a finite distance. Then we should be able to measure one spin state on one particle, and another spin state on the other. If so, we would know both states at the same time. Einstein believed that this will show that the limitation specified by Heisenberg's uncertainty principle can be overcome. The Heisenberg principle tells us that the more accurately we specify one of the parameters in the quantum state of a particle, such as momentum or spin, the less accurately can we specify its other parameters—for example, its location in space. When we fully define one parameter, the others become entirely blurred. The bottom line is that it is not possible to measure all parameters of the particle's quantum state at the same time. Einstein believed that this interdiction is not intrinsic to the reality of particle; it is due to our systems of observation and measurement. He suggested the EPR as a thought experiment to convince us that this is likely to be the case.

When experimental apparatus sophisticated enough to test the EPR experiment came online, it turned out that the Heisenberg uncertainty principle holds up. Indeed, it holds up under conditions that Heisenberg himself did not envisage: namely, over any finite distance.

Suppose that we measure the spin state of one of the particles—particle A—along some direction, let us say the *z*–axis (the permissible spin states are "up" or "down" along axes *x*, *y*, and *z*). Let us say that we find that this measurement shows the spin to be in the direction "up." Because the spins of the particles have to cancel each other, the *z*–axis spin of particle B must be "down." But if the particles are removed from each other, than this requirement should not hold. But it does! Every measurement on one particle yields a complementary outcome in the measurement on the distant other. The measurement of particle A has an instantaneous effect on B, causing the wavefunction of its spin to collapse into the complementary state.

It appears that the measurement on A doesn't just *reveal* an already established state of B: it actually *produces* that state. The state of B depends rigorously on the measurement on A. A signal seems to propagate from A to B, conveying information on what is measured on A. The effect on B is simultaneous with the measurement on A: if a signal is involved, it propagates with infinite velocity.

These findings indicate an unexpectedly close correlation between quanta, the smallest measurable entities of the physical world. They show that particles A and B remain connected—*instantly* connected—at any finite distance from each other.

How could this connection persist when the particles are separated? That a signal propagating from A and reaching B would create the connection has been considered: it is not the case. Since then scores of experiments have been carried out over ever greater distances. They testify that there is an instant interconnection between the particles, and it is not conveyed by any hitherto known effect propagating from one to the other.

Separation in space and time does not divide particles that originated in the same quantum state. It is not necessary that the particles should have actually originated in the same place and at the same time: it is enough that at one time and in one place they should have shared the same quantum state. Particles that at one time and one place occupied the same quantum state can be light years apart in space and thousands of years apart in time. Their connection is invariant with respect to distance and to time.

As early as 1935 Erwin Schrödinger noted that particles in the same quantum state do not have individually defined states: their states are intrinsically "entangled." The state of collective superposition applies to two or more properties of a single particle, the same as to a set of several particles. It is not the single particle or the single property of a particle that carries information on the quantum state, but the collective wavefunction of the system in which the particle is embedded.

Observations regarding the quantum world transcend the philosophical position known as "local state realism." Interaction in this world is not confined to the limits prescribed by classical physics, and even those of relativity theory. On the level of quanta, this is an entangled world—a world of universal nonlocal interaction.

Chapter Two

Anomalies in the Living World

At the initial phases of the quantum revolution it was generally believed that quantum phenomena are limited to the quantum world: at macroscopic scales the world is basically "classical." This, however, proved not to be the case. Nonlocal interaction surfaced also at macroscopic scales. The kind of coherence exhibited by systems at the scale of life provides indirect evidence for this: the kind and level of coherence that comes to light in the world of the living could hardly have come about in the absence of nonlocal interconnection among the parts and elements of living systems, and also among the systems themselves.

Nonlocality in the biosphere: (a) theory

More than half a century ago Erwin Schrödinger, convinced of the reality of what he termed "entanglement" in the quantum realm, suggested that this phenomenon need not be limited uniquely to the quantum domain. The kind of order manifested in living systems is not order of a mechanical kind: it is *dynamic* order. Dynamic order is not an order based on chance encounters among mechanically related parts, and it cannot arise by random collisions among individual molecules. It is an order based on ultra–rapid system–wide connection among all parts of the system, even among those that are not contiguous with each other.

Systemwide interconnection is the *sine qua non* of organic functioning. Organisms are carbon–based thermodynamic systems operating in a water–based medium. They maintain themselves within a flow of energy from the environment, compensating for the degradation of free energy; a necessary consequence of work performed within a system. Unless they constantly import the energy required to maintain their structure and function, their specific entropy increases, bringing them closer to the inert state of thermodynamic equilibrium. As Ilya Prigogine pointed out, life is physically possible only through the efficient storing and mobilization of free energy—negative entropy—in a dynamic state far from thermodynamic equilibrium.

Close cooperation based on a quasi–instant coordination among all parts of the organism is a precondition of maintaining the living system in its far–from–equilibrium condition. Even rare molecules need to find each other to produce the reactions that maintain the organism, and such molecules may not be in each other's immediate proximity. There would not be sufficient time for mutual recognition if it depended on a process of random sifting and mixing. If the organism is to access, mobilize, and store the energy required to maintain life's irreversible processes, the molecules must locate and interact with each other regardless of whether they are neighboring, or distant.

The thermodynamic imperative is to replace downgraded high–entropy energy with negentropic free energy, and if this imperative is to be satisfied, all parts of the organism must communicate with all other parts reliably and without delay. This kind of instant and multidimensional communication suggests that the kind of order that characterizes quantum systems also applies to systems in the domain of life. How this kind of order functions is becoming better understood.

According to biophysicist Mae–Wan Ho, living organisms maintain themselves through the superposition of two basic processes: a nondissipative cyclic process, for which the net entropy balances out to zero ($DS = 0$), and a dissipative, irreversible process for which entropy production is greater than zero ($DS > 0$).[4] The cyclic, nondissipative loop embraces almost all living components of the organism due to the ubiquity of the cycles that

4 Mae–Wan Ho *The Rainbow and the Worm: the Physics of Organisms*. 1993. World Scientific, Singapore.

constitute it. Its coupling with the irreversible energy throughput loop frees the living organism from immediate thermodynamic constraints. It makes use of the conservation principle of energy (the first law of thermodynamics) to offset the irreversible degradation of energy (the second law). This is possible in open systems that, as Schrödinger put it, "feed on negentropy."

For an organism to maintain itself by "feeding on negentropy," it must approximate in some respects a quantum system. In quantum systems molecular assemblies, whether neighboring or distant, resonate in phase since the same wavefunction applies to them. Attractive or repulsive forces are generated depending on the phase relations among the wavefunctions, and faster and slower reactions take place in regions where the wavefunctions coincide. Through such processes long–range correlations come about that are nonlinear, quasi–instant, heterogeneous, and multidimensional. Whilst molecular reactions at different points carry out the individual functions, the coordination of the functions occurs by means of a quasi–instant multidimensional transfer of information among the assemblies.

Quantum–processes within the organism are an anomaly for mainstream biology. Standard biological theory is based on definite, deterministic interactions; one part or element of the organism is said to determine the state or behavior of another part. But recent observations indicate that molecular interactions in the organism do not rigorously determine its functions and processes. Even the genetic variety of determinism (the doctrine that genes in the organism contain the full set of instructions for building the whole organism) is not adequate to the facts. It is clear that by means of complementary copies of messenger–RNA genes determine the amino acid sequence of protein molecules, but the resulting structures do not uniquely determine the structures and functions of the organism. The living system is not a mechanistic system where genetic instructions are rigorously transcribed into molecular structure. Rather, the organism is a highly integrated system where self–maintaining processes engage all levels simultaneously, from the microscopic to the molecular and macroscopic. Adjustments, responses and changes required for the maintenance of the system propagate in all directions at the same time, and are sensitively tuned to conditions in the organism's environment.

Cells in the organism are highly differentiated: some constitute building blocks, others play the role of enzymes, another class participates in cell–signaling, and some (the motor proteins) transform chemical energy into mechanical energy. Yet stem cells in the organism are genetically equivalent; they are only differentiated in the course of embryonic development. This process involves the activation or repression of different sets of genes. Genetic determinism claims that this is entirely genetic: sets of genes are responsible for it. This is not the case. Some basic developmental processes are either entirely outside genetic control, or are only indirectly affected by genes. Russian biophysicist Lev Beloussov suggested that the truth may be the reverse of genetic determinism: genes themselves could be obedient servants fulfilling powerful commands by the rest of the organism.[5]

Genetic determinism faces the so–called C–value paradox (where C stands for complexity and C–value denotes the size of the organism's haploid set of chromosomes, that is, the size of its DNA sequence), as well as the gene–number paradox (the paradox of gene redundancy). In regard to C–value, the empirical findings are contrary to expectations. If the information coded in the genome would provide a more–or–less complete description of the organism, the complexity of the phenome (the flesh–and–blood organism) and the complexity of the genome should be proportional: more complex organisms should have more complex genetic information. But genetic and phenotypic complexity are not correlated. The Human Genome Project identified less than 40,000 genes in the human genome, a surprisingly small amount. Even an amoeba has two hundred times more DNA per cell than a human being.

It is also puzzling that phylogenetically closely related species should have radically different genomes. The genome size of closely related rodents often varies by a factor of two, and the

5 Beloussov, Lev, 2002. "The formative powers of developing organisms," *What Is Life?* Hans–Peter Dürr, Fritz–Albert Popp and Wolfram Schommers, eds. New Jersey, London, Singapore: World Scientific

genome of the housefly is five times larger than the genome of the fruit fly. At the same time some phylogenetically distant organisms have similar genetic structure. Given these discrepancies, it is difficult to see how the structure of the genome would determine the structure of the phenome. One and the same gene can produce different proteins, and different genes can produce one and the same protein. The complex and coherent functioning of the living organism cannot be explained uniquely, or even primarily, in reference to genetic instruction governing molecular interaction.

The living state is maintained not by genetic, but by *epigenetic* regulation. A system of regulation within the organism turns genes "on" and "off"—activating and de–activating the genetic endowment of the species to meet the survival needs of the organism. Experiments show that epigenetic self–regulation has long–standing effects: it can even be handed down to succeeding generations.

The organism is intrinsically coherent, with ultrasensitive connections binding its diverse parts into functional unity. Signals even below the threshold of biochemical sensitivity produce measurable effect. The human ear, for example, can pick up sounds below the level of thermal noise, and the cat's eye can respond to a single photon. Bioelectrodynamics research shows that extremely low–intensity electromagnetic fields, with energy below the physical thermal noise limit, can produce observable effect.

The new findings regarding the supersensitivity of the organism, together with the puzzles facing molecular and genetic determinism, corroborate Hans Fröhlich's bold hypothesis that all parts of the living system create fields at various frequencies that infuse the organism and radiate into the environment. The specific resonance frequency of each molecule, cell, tissue, and organ creates long–range phase correlations similar to those that occur in superfluidity and superconductivity.

Even photosynthesis, the basis of all life on the planet, relies on quantum effects. Gregory Engel and collaborators found evidence that long–lived electronic quantum coherence plays an important part in photosynthesis. The quantum coherence–based wavelike characteristic of the energy transfer explains the extreme efficiency of the photosynthetic process by allowing the molecular complexes to sample vast areas if phase space and find the most efficient path.[6]

According to Mae–Wan Ho, all living systems are permeated by quantum waves. The biological organism, the phenome, is where the wavefunctions are the most dense.

Physics experiments underscore the validity of this hypothesis. It appears that living tissue constitutes so–called Bose–Einstein condensates. These condensates were first postulated in 1924, but it was only seven decades later, in 1995, that experiments could demonstrate their presence in the organism. Eric A. Cornell, Wolfgang Ketterle and Carl E. Wieman received the 2001 Nobel Prize "for the achievement of Bose–Einstein condensation in dilute gases of alkali atoms, and for early fundamental studies of the properties of the condensates." They have shown that supercooled aggregates of matter—in their experiments rubidium or sodium atoms were used—behave as nonlocal waves. They penetrate throughout the condensate and form interference patterns.

Bose–Einstein condensates have been found in living tissue at the temperature of warm–blooded animals. They are a hybrid form of matter that is neither solid nor liquid but "liquid crystalline." Some tissues of the organism (for example, connecting tissues and cell membranes, and the collagen in bones) prove to be liquid crystals in part or in whole.

Information within Bose–Einstein condensates is transferred instantly, producing the kind of coherence previously associated only with lasers and quantum systems. These effects were thought to be limited to the quantum domain; it was thought that at ordinary temperatures Brownian movement creates decoherence. But this is not necessarily the case. Recent theories, advanced among others by Kitaev, Pitkanen, and Frecska and Luna, suggest that there can be specifically organized networks of quantum particles—for example, networks where the particles are "woven" or "braided"—that

6 Engel, Gregory S., Tessa R. Calhoun. Elizabeth L. Read, Tae–Kyu Ahn, et al. 2007 Evidence for wavelike energy transfer through quantum coherence in photosynthetic systems. *Nature* 446, 782–786 (12 April).

are sufficiently robust to maintain coherence at biological temperatures.[7] While at such temperatures quantum systems suffer "heat–decoherence," in networks of woven or braided quanta coherence is conserved. As Parsons put it, "braiding is robust: just as a passing gust of wind may ruffle your shoelaces but won't untie them, data stored on a quantum braid can survive all kinds of disturbance."

Nonlocality in the biosphere: (b) evidence

We now turn to evidence for nonlocal connection in the living world. Evidence, both observational and experimental, comes from a variety of sources, some of them unexpected and on first sight surprising.

Evidence for nonlocal connection between parts of the organism—In the 1960s lie–detector expert Cleve Backster placed the electrodes of his lie–detector (a polygraph) on the leaves of a plant in his office. To his surprise, he found that the instrument registered reactions by the plant that correlated with his own experiences. For example, the polygraph revealed marked deviations in the electrical resistance of the plant at the very instant when Backster nearly had an accident on the street below. The correlation was maintained even when the leaf was detached from the plant and trimmed to electrode size, or shredded and redistributed between the surfaces of the electrodes.

These strange findings have been replicated in subsequent experiments. Ben Bending of the University of California at Los Angeles reported that his results "support the claim that plants have some sort of perceptual faculties that allow them to sense human emotion."* Although the signals manifest electrically, the connection may not be electromagnetic, given that Faraday cages and lead shielding fail to block the signals. According to Bending this suggest the possibility of a nonlocal connection between plants and their caretakers.

7 Frecska, Ede and Luna, Luis Eduardo, 2006. "Neuro–Ontological Interpretation of Spiritual Experiences," *Neuropsychopharmacologia Hungarica* 8 (3)

* Bending, B. W. Plant Sensitivity to Spontaneous Human Emotion. Poster session presented at: Toward a Science of Consciousness; 2012 April 10–14; Tucson, AZ

Evidence for nonlocal connection among different parts of the same organism was uncovered when Backster took oral leukocytes (white cells) from the mouth of test subjects and moved them to locations ranging from 5 yards to over 8 miles distant. He monitored the cells' net electrical potentials and fed the signals to a chart drive–unit that provided a continuous recording of the changes. He then stimulated his subjects with visual images designed to evoke an emotional response and observed the variations in the cells' electrical potentials. It turned out that the changes in the potentials were correlated in time, amplitude, as well as duration with the emotional responses of the subjects.[8]

In one experiment a young man was handed an issue of *Playboy* magazine and allowed to peruse its pages. When he came to the centerfold photo, which was a nude picture of actress Bo Derek, his EEG manifested an emotional response that lasted during the whole time he viewed the picture. Changes in the electrical potentials of his distant cells mirrored the changes in his emotional state. When he closed the magazine the values of his responses returned to average. And when he decided to reach for the magazine for another look, the same reaction was repeated also in the cells.

A similar correlation was noted when a retired U.S. Navy gunner was tested who had been at Pearl Harbor during the Japanese attack. When watching a TV program entitled "The World At War," the gunner did not react to the downing of enemy aircraft by naval gunfire. He did react, however, when the downing occurred immediately following a facial close–up of a naval gunner in action. At that point—when he seemed to have projected his own wartime experiences into the scene—his cells, located at a distance of 8 miles, showed a reaction that was precisely correlated with his own.

Persistent nonlocal connection between cells removed from an organism and the host organism was discovered also in an unrelated set of experiments by a group of medical doctors in the United Kingdom, members of the Lawrence Society for Integral Medicine.[9] They sought to access what they call the "psi field," a nonlocal

8 Backster, Cleve, 1968. "Evidence of a Primary Perception at the Cellular Level in Plant Life" *Int. Journal of Parapsychology* 10.4

9 *Psionic Medicine*, Journal of the Psionic Medical Society and The Institute of Psionic Medicine. 2000.Vol. XVI.

field that is said to envelop the organism (the group was originally called "The Psionic Medical Society"). To access information in the psi–field, the physicians make use of a sophisticated form of medical dowsing. They found that they could diagnose their patients at any distance; all they needed was a so–called "witness," which could be any protein sample from the body of the patient, such as a strand of hair or drop of blood. They could perform the diagnosis by observing the movement of a pendulum over a specially designed chart. In thousands of cases, decoding the pendulum's movement produced a correct diagnosis of the patients' condition.

The cells that make up the witness can be analyzed repeatedly, at any time and at any distance from the patient. The information they yield reflects the patient's state of health at the time that analysis is carried out, and not at the time the cells were removed from the body of the patient. This suggests that it is not the cells that convey the information (because if they did, the information would reflect the condition of the patient at the time the cells were removed). The cells remain nonlocally connected with the body of the patient and reflect the latter's physical condition whenever the tests are carried out.

These findings, though surprising on first sight, stand to reason: superfast, distance–independent connection among the parts of the organism is essential if the organism is to maintain the coherence it needs to sustain itself in a physically unstable state far from thermal and chemical equilibrium. The staggering number of physical and chemical reactions in the organism cannot be adequately coordinated by narrow–band and slow neural and biochemical signal–transmission. Only the nonlocal "entanglement" of the organism's cellular and subcellular components can ensure a sufficiently rapid flow of multidimensional information to maintain the system in its environment.

Evidence for nonlocal connection between different individuals—Nonlocal connections are not limited to the parts of the same organism; they also surface between discrete organisms. This comes to light in experiments with individuals located at distances that preclude physical and physiological interaction among them.

A set of experiments using functional magnetic resonance imaging (fMRI) tested the connection between the cerebral activity of individuals who were not in any known form of communication with each other.[10] Jeanne Achterberg of the Saybrook Institute asked eleven healers to choose persons with whom they felt an empathic connection. The selected subjects were placed in an MRI scanner isolated from sensory contact with the healers. The healers sent energy, prayer, or good intentions—so–called distant intentionality—at random intervals that were unknown to both the healers and the subjects. In analyzing the results Achterberg found significant differences between the "send" and "no send" (control) periods in the activity of various parts of the subjects' brain, namely the anterior and middle cingulate areas, precuneus, and the frontal areas. The probability that these differences would have occurred by chance was calculated as one in 10,000.

Tests regarding the effects of conscious intention offer further evidence for nonlocal connection between individuals. It has been long known that the focused intention of one person can affect the bodily state of another. This was confirmed by anthropologists researching what is known as "sympathetic magic" in traditional cultures. In his famous study *The Golden Bough*, Sir James Frazer noted that Native American shamans practicing voodoo would draw the figure of a person in sand, ashes, or clay, and then prick it with a sharp stick or do it some other injury. The corresponding injury was said to be inflicted on the person the figure represented. Anthropologists found that the targeted person often fell ill, became lethargic, and would sometimes die. Consciousness researcher Dean Radin and his collaborators at the University of Nevada tested the positive variant of this effect under laboratory conditions.

In Radin's experiments the subjects created a small doll in their own image, and provided various objects (pictures, jewelry, an autobiography, and personally meaningful tokens) to "represent" them. They also gave a list of what makes them feel nurtured and comfortable. These and the accompanying information were used by the healers to create a sympathetic connection to the subjects.

10 Achterberg, J., K. Cooke T. Richards, L. Standish, L. Kozak, and J. Lake, 2005. "Evidence for Correlations between Distant Intentionality and Brain Function in Recipients: A Functional Magnetic Resonance Imaging Analysis," Journal of Alternative and Complementary Med. 11,6.

The latter were wired to monitor the activity of their autonomous nervous system—electrodermal activity, heart rate, and blood pulse volume—while the healer was in an acoustically and electromagnetically shielded room in an adjacent building. The healer placed the doll or the other objects provided by the subjects on a table and concentrated on them while sending randomly sequenced "nurturing" (active healing) and "rest" messages to the subjects.

In the experiments the electrodermal activity and heart rate of the subjects were significantly different during the active nurturing periods and during the rest periods, while blood pulse volume was significant for a few seconds during the nurturing period. Both heart rate and blood flow indicated a "relaxation response"—which made sense since the healer was attempting to "nurture" the subject via the doll. On the other hand, a higher rate of electrodermal activity showed that the subjects' autonomic nervous systems was becoming aroused. Why this should be so was puzzling until the experimenters realized that the healers nurtured the subjects by rubbing the shoulders of the dolls that represented them, or stroked their hair and face. For the subjects this worked like a "remote massage." Radin concluded that the local actions and thoughts of the healer are mimicked in the distant subject almost as if healer and subject were next to each other.

Radin's findings have been corroborated by William Braud and Marilyn Schlitz in hundreds of experiments over more than a decade. The experiments tested the impact of the mental imagery of senders on the physiology of receivers. The effects proved similar to those produced by the subjects' own mental processes on their own body. "Telesomatic" action by a distant person proved to be nearly as effective as "psychosomatic" action by the subjects themselves.

Experiments that demonstrate the effectiveness of distant healing were pioneered by cardiologist Randolph Byrd, a former professor at the University of California at Berkeley. His ten–month computer–assisted study concerned the effects of intention on patients admitted to the coronary care unit at San Francisco General Hospital.[11] Byrd formed a group of experimenters made up of

11 Byrd, Randolph, 1988. "Positive therapeutic effects of intercessory prayer in a coronary care population," *Southern Medical Journal* 81:7

ordinary people whose only common characteristic was a habit of regular prayer in Catholic or Protestant congregations around the country. The selected people were asked to pray for the recovery of a group of 192 patients; another set of 210 patients, for whom nobody prayed in the experiment, made up the control group. Rigid criteria were used: the selection was randomized and the experiment was carried out double–blind, with neither the patients nor the nurses and doctors knowing which patients belonged to which group. The experimenters were given the names of the patients, some information about their heart condition, and were asked to pray for them every day. They were not told anything further. Since each experimenter could pray for several patients, each patient had between five and seven people praying for him or her.

The results were statistically significant. The prayed–for group was five times less likely than the control group to require antibiotics (3 versus 16 patients); it was three times less likely to develop pulmonary edema (6 compared to 18 patients); none in the prayed–for group required endotracheal intubation (while 12 patients in the control group did); and fewer patients died in the former than in the latter group (though this particular result was statistically not significant). It did not matter how close or far the patients were to those who prayed for them, nor what type of praying was practiced—only the fact of concentrated and repeated prayer seemed to have counted, without regard to whom the prayer was addressed and where it took place.

In the form of alternative medicine physician Larry Dossey calls "Era III nonlocal medicine," nonlocal effects are used systematically for healing.[12] For example, a sensitive is asked to concentrate on a given patient from a remote location. As shown by the practice of various healers, it is enough to give the name and date of birth of the patient. Neurosurgeon Norman Shealy often telephoned this information from his office in Missouri to clairvoyant diagnostician Carolyn Myss in New Hampshire. She diagnosed the cases and sent the results to Dr. Shealy. The latter found that in the first one hundred cases her diagnosis was 93 percent correct.

12 Dossey, Larry, *Recovering the Soul: A Scientific and Spiritual Search,* 1989. New York: Bantam.

Evidence for nonlocal connection between organisms and their environment—There are nonlocal connections among parts of the same organism, and nonlocal connections exist also between different organisms. Moreover it appears that there are nonlocal connections also between organisms and their milieu.

The evidence for this proposition is indirect. It concerns the extremely low probability that Darwin's classical tenet would hold true—that living species would evolve by chance mutations in their genome.

The "synthetic theory," the current version of Darwin's original concept, claims that chance mutations in the genome create the kind of modifications in the information pool of a species that transform an unviable species into a viable one. This, however, turns out to be statistically extremely improbable.

The oldest rocks date from about four billion years before our time, while the earliest and already highly complex forms of life (blue–green algae and bacteria) are over 3.5 billion years old. Half a billion years seems like a long time, but it may not be long enough to explain how complex species could have evolved. The assembly even of a primitive prokaryote involves building a double helix of DNA consisting of some 100,000 nucleotides, with each nucleotide containing an exact arrangement of thirty to fifty atoms, together with a bilayered skin and the proteins that enable the cell to take in food. This construction requires an entire series of reactions, finely coordinated with each other. If living species had relied on random variation in an isolated genome, the level of complexity we observe in the domains of life is not likely to have been achieved in the approximately half a billion years that was available for it.

The probability that random chances would produce complex species within realistic timeframes is reduced by additional considerations. It is not enough for mutations to produce one or a few positive changes in the genetic pool of a species; if the changes are to be viable, they must involve the entire genome. The evolution of feathers, for example, doesn't produce a reptile that can fly: radical changes in musculature and bone structure are also required, along with faster metabolism to power sustained flight. These changes involve complex processes: at least nine varieties of genetic rearrangements are known (transposition, gene duplication, exon shuffling, point mutation, chromosomal rearrangement, recombination, crossing–over, pelitropic mutation, and polyploidy). Many of them are interrelated. It is not very likely that these rearrangements, whether they occur singly or in combination, would produce new species from the old by chance variations of the genome. A random mutation is not likely to result in evolutionary advantage; it is likely to make the species less rather than more fit to survive. And in that case it would ultimately be eliminated by natural selection. Yet many species turned out to be viable, and their evolution was far more rapid than the search–space of chance variation would allow.

The synthetic theory is further challenged by evidence that the genome is not fully isolated from the phenome, and hence it cannot mutate in isolation. But even the "fluid genome," responsive to inputs from the phenome, fails to explain the precise, rapid, and highly focused macromutations called for to transform the genetic information of an extinction–threatened species into information that can code a new and viable species. There must be sufficient sensitivity on the part of the genome–phenome system to changes in its milieu in order to produce the necessary mutations. This level of sensitivity exceeds the currently recognized interactions between organisms and environments.

Thus it is extremely improbable that viable species would have emerged through random mutations in the genome. According to mathematical physicist Sir Fred Hoyle, the probability in question is about the same as that of a hurricane blowing through a scrap-yard assembling a working airplane.

As early as 1937, biologist Theodosius Dobzhansky said that the birth of new species by genetic mutation would be impossible in reality, even if the birth of new species would occur on a "quasi–geological scale." However, new species appear much faster than that. In their theory of "punctuated equilibrium" Stephen Jay Gould and Niles Eldredge noted that the populations that are most likely to mutate are peripherally isolated and relatively small. Changes in their genome are fast and yet precise, often taking no more than five to ten thousand years. This makes the geological time of species evolution into an insignificantly short time—an evolutionary instant.

Chapter Three

Anomalies in the Universe

The basic elements of the universe are intrinsically nonlocal, and nonlocal connections surface in the living world as well. And it appears that nonlocality extends also to the largest dimensions: to the scale of the universe itself. This, however, is an anomaly for the dominant model in cosmology.

The standard model

According to the standard model, a single event created the world, a nonrecurring and unexplained singularity known as the Big Bang.

The universe is to have originated in an explosive instability 13.75 ± 0.13 billion years ago. A region of cosmic pre–space had exploded, creating a fireball of staggering heat and density. In the first few milliseconds it synthesized all the matter that populates space and time. The particle–antiparticle pairs that emerged from the vacuum collided with and annihilated each other; and the one billionth of the originally created particles that survived the collisions (the tiny excess of particles over antiparticles) constitutes matter in the universe.

After about four hundred thousand years photons decoupled from the radiation field of the primordial fireball: space became transparent, and clumps of matter established themselves as distinct entities. Due to gravitational attraction these clumps condensed into stars and stellar systems, and created gigantic swirls that, after about a billion years, became galaxies.

The universe that evolved in the wake of the primal explosion turned out to be precisely balanced between contraction and expansion. The computer analysis of some three hundred million observations first made through NASA's Cosmic Background Explorer Satellite (COBE) in 1991 provided remarkable confirmation. Detailed measurements of the cosmic microwave background radiation—the presumed remnant of the Big Bang—indicate that the variations are not distortions caused by radiation from stellar bodies but are the remnants of minute fluctuations in the cosmic fireball when it was less than one trillionth of a second old. In April 1992, a team of astrophysicists led by George Smoot announced that they had mapped the intensity of the radiation from the primal fireball and found tiny variations that, under the action of gravitation, acted as the seeds from which the macrostructures of the universe had evolved. [13]

Measurements of the background radiation disclose the matter–density of the universe, that is, the number of particles that were created in the primal explosion and were not annihilated in the particle–antiparticle collisions that followed it. If the surviving particles make for a matter–density above a certain number (estimated at 5×10^{-26} g/cm^3), the gravitational pull associated with the total amount of matter will ultimately exceed the inertial force generated by the primal explosion, and the universe is closed: it will collapse back on itself. If matter–density is below that number, expansion will continue to dominate gravitation. Then the universe is "open": it will expand indefinitely. However, if matter–density is precisely at the critical value, the forces of expansion and contraction will balance each other and the universe is "flat." It will remain balanced at the razor's edge between the opposing forces of expansion and contraction.

Recent findings show that the universe created in the wake of the primal explosion was fine–tuned to the staggering precision of one part in 10^{50}. A deviation even of that minute order would have produced either an infinitely expanding open universe, or a recollapsing closed universe.

Sophisticated measurements also disclose that not only is the matter–density of the universe precisely tuned for balance between expansion and contraction; also the dimensions of the

13 Smoot, George; Keay Davidson (1994). *Wrinkles in Time.* William Morrow & Company.

existing particles are finely tuned to the macro–parameters of the universe. Applying Einstein's celebrated relation between mass and energy ($E = mc^2$), the size of the electron (which is $ro = 6 \cdot 10^{-15}$ meters) turns out to be a consequence of the number of electrons in the universe (the latter as given by Eddington's number, approximately 2×10^{79} in the Hubble universe of $R = 10^{26}$ meters). Moreover the physical constants of the universe are also finely tuned to each other. Physicist Dezsõ Sarkadi discovered that a simple exponential rule ($Q = 2/9$) is maintained when the relation of one constant to the other is quantified. [14]

Menas Kafatos and collaborators came up with further relations between the macro–parameters of the universe. Scale invariant relations appear between the mass constituted by the total number of particles in the universe, the gravitational constant, as well as the charge of the electron, Planck's constant, and the speed of light. Among other things, all lengths turn out to be proportional to the scale of the universe. This suggests a staggeringly high level of coherence among the constants of the cosmos. Kafatos states unequivocally that the entire universe, being instantly interconnected, is nonlocal.[15]

Coherence, and hence nonlocality, comes to light also in regard to the processes that lead to the evolution of life in the universe. The basic forces and constants are precisely tuned to create conditions that permit complex systems to emerge. A minute difference in the strength of the electromagnetic field relative to the gravitational field would have prevented the evolution of such systems, since hot and stable stars such as our Sun could not have come about. If the difference between the mass of the neutron and the proton would not be precisely twice the mass of the electron, no substantial chemical reactions could take place. And if the electric charge of electrons and protons did not balance precisely, all configurations of matter would be unstable and the universe would consist only of radiation and a relatively uniform mixture of gases.

---

14 Sarkadi, Dezsõ and László Bodonyi, 1999. "Gravity between commensurable masses." Private Research Center of Fundamental Physics, *Magyar Energetika*, 7:2

15 Menas, Kafatos and Robert Nadeau, *The Non–Local Universe: the New Physics and Matters of the Mind.* Oxford University Press, 1999.

Yet in this universe the gravity constant ($G = 6.67 \times 10^{-11}$) is precisely such that stars can form and shine long enough to allow the evolution of the stellar structure of galaxies, including systems with planets that orbit hot and stable stars. If *G* would be smaller, particles would not compress sufficiently to achieve the temperature and the density needed to ignite hydrogen: stars like our Sun would have remained in a gaseous state. If on the other hand *G* were larger, stars would have formed but would burn faster and endure for a shorter time, making it unlikely that life could evolve on the planets orbiting them.

Likewise, if the Planck constant ($h = 6.63 \times 10^{-31}$ Jsec) would be even minutely other than it is, carbon–producing nuclear reactions could not occur in stars—and consequently complex systems based on carbon–bonding could not arise on some of their planets. But given the actual value of *G* and *h*, and of an entire array of universal constants (including the velocity of light, the size and mass of the electron, and the relationships between the size of the proton and the nucleus), complex systems could in fact arise on some planets around active stars.

A further feature of coherence in the universe is the uniformity of the background radiation. The microwave background radiation, emitted when the universe was about a hundred thousand years old, is known to be isotropic, the same in all directions. This calls for an explanation. Physical signals could not have tuned the various regions of the expanding universe, because when the background radiation was emitted the universe's opposite sides were already ten million light years apart—and light could only have traveled one hundred thousand light years in that time. Although they were not connected by classical forces and signals beyond the first few microseconds after the explosion that created our universe, even distant galaxies and other macrostructures evolve in a remarkably uniform manner.

In 1997 cosmologist Alan Guth proposed the theory of inflation to account for this "horizon problem."[16] According to this theory, in the span of the 10^{-33} sec Planck–time following the birth of the universe, space expanded at a rate faster than light. This did

16 Guth, Alan H., 1997. *The Inflationary Universe: The Quest for a New Theory of Cosmic Origins*. California: Perseus.

not violate general relativity, since it was not matter that moved at a supraluminal velocity but space—matter stood still relative to space. During the initial Planck–time all parts of the universe were in contact, sharing the same density and temperature. Then, as the universe expanded, some parts moved beyond the range of physical contact and evolved independently. They could evolve uniformly, since they were connected immediately after the birth of the universe.

The presence of life

For long, life in the universe was considered to be a chance occurrence, based on the lucky coincidence that a set of in–themselves improbable conditions would have come together. It was known that for life to appear not only the basic constants and parameters of the universe need to be finely adjusted, further conditions need also to be met. There must be a planet with the correct mass at the right distance from a main sequence G2 dwarf star; the planet needs to occupy a nearly circular orbit; it must have an oxygen/nitrogen rich atmosphere, a large moon and a moderate rate of rotation; it must be at the right distance from the center of the galaxy and have liquid water on its surface, and it must have a correct ratio between water and land–mass. Last but not least, it must have protection from asteroids through giant gas planets in its solar system.

It was held that the requirement for such improbably coinciding factors makes life an exceptional event in the universe. But this view was placed in question by the surprising finding published in October of 2011. A team of researchers headed by Sun Kwok and Yong Zhang at the University of Hong Kong reported that organic molecules, the basic building blocks of life, are created in stars.[17] Some 130 such molecules have been found so far, including glycine, an amino acid, and ethylene glycol, the compound associated with the formation of the sugar molecules necessary for life.

17 Sun Kwok, *Organic Matter in the Universe.* Wiley 2011; Sun Kwok and Young Zhan, Astronomers Discover Complex Organic Matter Exists throughout the Universe. *Science Daily,* October 26, 2011.

Working with NASA's Spitzer Space Telescope, they found that water, methanol, and carbon dioxide coat dust particles around stars 420 light–years away in the constellation Taurus. These substances show up in interstellar dust clouds and dusty planet–forming discs around stars.

It appears that at various stages of their evolution active stars eject organic compounds into interstellar space, distributing the molecules over vast regions. NASA noted that, while such materials have been found elsewhere, "this is the first time they were seen unambiguously in the dust making up planet–forming gases."

This observation is an anomaly for mainstream astrophysics. "Our work has shown that stars have no problem making complex organic compounds under near–vacuum conditions," Professor Kwok noted. "Theoretically, this is impossible, but observationally we can see it happening."[18]

This observation, though anomalous, is of great significance. It suggests that life is not a random or exceptional event in the universe. The very processes of stellar evolution create the basis for the evolution of biological systems. Only the higher forms of life—complex biochemical systems capable of metabolism and reproduction—require the highly specific combination of conditions that are likely to be relatively rare in the universe.

Coherence and nonlocality in the universe: is there an explanation?

It is highly unlikely that the coherence observed in the universe would be the result of mere chance. While it is true that the theory of large numbers shows that in a large number of tries even otherwise improbable outcomes have a reasonable probability of coming about, the number of tries needed to reach a significant probability that a universe such as ours would come about is extremely high. According to calculations by Roger Penrose, $10^{10^{123}}$ universes are physically possible; they are the effective "search–space"

18 Sun Kwok, interview quoted on the Internet-based, 'Science Daily' "Astronomers Discover Complex Organic Matter Exist throughout the Universe". http://www.sciencedaily.com/releases/2011/10/111026143721.htm

for the selection of a universe. Yet all but a small handful of this staggeringly large number of universes is biologically sterile, incapable of giving rise to the level of complexity believed to be a precondition of life.

Could our universe have been born in the context of these myriad alternatives through anything other than a random selection? This calls for showing first, that a large number of universes could indeed coexist, and second, that the selection from among them is not random. The first of these requirements is being met by a growing number of cosmological models, and meeting the latter has become significantly more than an abstract possibility.

So–called multicyclic universe models claim that a large number of universes could co–exist, or come successively into being, in the underlying meta–universe or "metaverse."

One of the first multicyclic models to be put forward was the work of Princeton physicist John Wheeler. In his model, advanced more than three decades ago, the expansion of our universe comes to an end, and the universe collapses back on itself. Following this "Big Crunch" a new expansionary phase gets under way. In the quantum uncertainties that dominate the supercrunched state nearly infinite possibilities exist for the emergence of a new universe.

New universes could also be created inside black holes. The extreme high densities of these space–time regions represent recurrent singularities where the laws of physics do not apply. Stephen Hawking and Alan Guth suggested that under these conditions the black hole's region of space–time detaches itself from the rest and expands to create a universe of its own. Thus the black hole of one universe may be the "white hole" of another: it is the "Bang" that creates it.

Still another multicyclic cosmology was elaborated by Ilya Prigogine, J. Geheniau, E. Gunzig, and P. Nardone.[19] Their theory suggests that major matter–creating bursts occur from time to time. The large–scale geometry of space–time creates a reservoir of "negative energy" (which is the energy required to lift a body

19 Prigogine, Ilya, J. Geheniau, E. Gunzig, and P. Nardone, 1988. "Thermodynamics of cosmological matter creation," *Proceedings of the National Academy of Sciences* USA, 85.

away from the direction of its gravitational pull); from this reservoir, gravitating matter extracts positive energy. Thus gravitation is at the root of the ongoing synthesis of matter: it produces a perpetual matter–creating mill. The more particles are generated, the more negative energy is produced and then transferred as positive energy to the synthesis of still more particles. Given that the cosmic vacuum is unstable in the presence of gravitational interaction, matter and vacuum form a self–generating feedback loop. A critical matter–triggered instability causes the vacuum to transit to the inflationary mode, and that mode marks the beginning of a new era of matter synthesis—a new universe.

Other cosmologies postulate the continuous creation of matter as well, among them the QSSC (Quasi–Steady State Cosmology) advanced by Fred Hoyle with George Burbidge and J. V. Narlikar.[20] Hoyle's original model called for a continuous, relatively linear creation of matter in space. Baby universes are periodically created in bursts similar to that which brought forth our own universe. These "matter–creating events" are interspersed throughout the universe. As the original version of the theory encountered empirical discrepancies, it was replaced by a version that called for matter–creation preferentially in regions of high matter–density where gravity is strongly negative. The revised theory suggests that the most recent matter–creating burst occurred about fourteen billion years ago, in good agreement with independent estimates of the age of the universe.

A recent multicyclic cosmology is the work of Paul J. Steinhardt of Princeton and Neil Turok of Cambridge.[21] Their model accounts for all the facts accounted for by the standard model, and it also explains an observation that is an anomaly for the latter: the accelerated expansion of distant galaxies. According to Steinhardt and Turok, the universe undergoes an endless sequence of cosmic epochs, with each epoch beginning with a "Bang" and ending with a "Crunch." Each cycle goes through

---

20 Hoyle, Fred, 1983. *The Intelligent Universe*. London: Michael Joseph. G. Burbidge and J. V. Narlikar, "A quasi–steady state cosmology model with creation of matter", *The Astropysical Journal* 410 (20 June 1993)

21 Steinhardt, Paul J. and Neil Turok, 2002. "A cyclic model of the universe," *Science*, vol 296, 1436–1439.

a period of gradual and then accelerating expansion, followed by reversal and contraction. They estimate that at present we are about fourteen billion years into the current cycle, at the beginning of a trillion–year period of continuous, and continuously accelerating, expansion. Ultimately space becomes homogeneous and flat, and a new cycle gets under way.

Leonard Susskind's model suggests that the staggering number of universes implied by M–theory, a recent version of string theory, is not a flaw in that theory but a profound insight into the nature of reality: each solution of M–theory's equations corresponds to a real universe, with all its laws and constants. The full range of universes governed by all the possible laws is the "landscape," and the collection of all the universes described by these laws is the meta–universe or metaverse.[22]

The same idea is at the core of Andrei Linde's version of inflation theory.[23] According to Linde, the super–rapid explosion that accompanied the birth of our universe was reticular, made up of several individual regions. What we know as the Big Bang had distinct regions, much like a soap bubble in which smaller bubbles are stuck together. As such a bubble is blown up, the smaller bubbles separate, each forming a distinct bubble of its own. The bubble universes percolate outward and follow their own evolutionary path. Each universe produces its own physical constants, and these may be very different from one another. For example, in some universes gravity may be so strong that material structures re–collapse almost instantly; in others, gravity may be so weak that no stars will form. Our bubble universe happens to provide just the right conditions for the evolution of complex systems, and hence of life.

We now turn to the second requirement for understanding the origins of the coherence found in our universe. It appears that the selection of our remarkably coherent universe from the myriad alternative universes may not have been a matter of simple serendipity. A model developed by a team of astrophysicists led

22 Susskind, Leonard, The Cosmic Landscape: String theory and the Illusion of Intelligent Design. Little Brown & Co. 2005.

23 Linde, A.D., *Particle Physics and Inflationary Cosmology*, Harwood Academic Publishers, Chur, Switzerland 1990.

by Abhay Ashtekar at the Institute for Gravitational Physics and Geometry at Penn State University gives substance to this possibility.[24] Their model permits the calculation of the state of the universe (actually, the metaverse) prior to the "Bang" that gave rise to it. This is a remarkable achievement, for the equations of general relativity do not permit going back to the time immediately after that cosmic inflation: matter was so dense then that the equations of general relativity did not hold. Quantum equations can penetrate further, but they were not available to Einstein. The mathematics of loop quantum gravity furnish the needed equations and permit "retrodicting" the conditions that reigned in the metaverse not just immediately after the birthing of our universe, but also prior to it.

In loop quantum gravity cosmology the fabric of space is a weave composed of one–dimensional quantum threads. Einstein's four–dimensional continuum is only an approximation; the geometry of spacetime has a discrete "atomic" structure. Before and during the primal explosion this fabric was torn apart, making the granular structure of space dominant. Gravity shifted from a force of attraction to a force of repulsion, and this led to the creation of the spacetime and the matter–content of our particular universe.

The simulations of loop quantum gravity indicate that prior to the birth of our universe there was a contracting universe with physical characteristics similar to ours. Ashtekar and collaborators were surprised at this finding and kept repeating the simulations with different parameter values. But the finding held up. It appears that our universe was not born in the singularity known as the Big Bang, and it will not end in the singularity of a Big Crunch. *The* universe is a metaverse that produces successive universe–cycles. The events viewed as Big Bang and Big Crunch are phase transitions in the metaverse, points where spacetime shrinks to quantum dimensions. The matter–component of the prior universe "evaporates" in black holes and is

24 A. Ashtekar, R.C. Cohen, D. Howard, J. Renn, S. Sarkar and A. Shimony (Editors), 2003. *Revisiting the Foundations of Relativistic Physics*, Festschrift in honor of John Stachel, Boston Studies in Philosophy of Science, Volume 234, Kluwer Academic.

re–born again in the superfast expansion that follows. Instead of a Big Bang leading ultimately to a Big Crunch, we have recurring Big Bounces.

At first, the consistency of the physical parameters of the successive universes—the cycles of the metaverse—was a puzzle. Given that at the transition from one cycle to the next the structure of spacetime is torn apart, it is not probable that the physics of the pre–existing phase should be reproduced in the next. Yet this appears to be the case. Alejandro Corichi of the National University of Mexico came up with equations that underscore this possibility.[25] His theory shows that a "semiclassical state" on one side of the Bounce peaks on a pair of canonically conjugate variables, and these strongly bound fluctuations on the other side. The change in fluctuations on the two sides is insignificantly small, (10–56) even for a universe of 1 megaparsec, and becomes smaller still for larger universes.

In their further development these theories could furnish good reason to maintain that the explosion that marks the birth of a universe does not produce cosmic amnesia. The metaverse may maintain nearly total recall across its cycles. Thus whatever coherence has been evolved in one universe will be conserved over the transition to the next, and will then be transferred to the one after that.

Our nonlocal universe may be far more than a chance occurrence in the cosmos. It could be the result of the coherent evolution of successive cycles in a deeper and more enduring, possibly infinite, metaverse.

---

25 Corichi, Alejandro, Black holes in loop quantum gravity. 2007. Morelia, Mexico ICGC'07, IUCAA.

PART II

INTRODUCING THE AKASHA PARADIGM

The quantum world, the living world, and the universe itself, are highly and in the context of contemporary science surprisingly, coherent. Coherent systems relate coherently to each other and together form a supercoherent system—a universe constituted of interacting and evolving systems within systems.

Coherence is not an exceptional, rare, or questionable phenomenon. It is quasi universal, coming to light in repeated and controlled experiments across the full range of scale in nature from the microworld of the quantum, through the mesoscale world of life, to the macroscale world of the universe.

The coherence of the observed phenomena, and the supercoherence of the systems constituted of the coherent phenomena, suggest nonlocal interaction within and among the systems. Nonlocality, however, is anomalous for the presently dominant local–state paradigm. Regardless of whether a bona fide explanation is found for the origins of our universe's coherence and thus for the presence of nonlocal interaction among its components, the coherence and nonlocality we find is an observational fact and it needs to be accounted for. Doing so calls for shifting to a nonlocal–state paradigm.

The chapters that follow outline the basic assumptions of the nonlocal–state paradigm we call "Akasha" in recognition of an ancient insight.

Chapter Four

Fields

As Einstein noted, the aim of science is to discover the scheme that can tie together the observed facts consistently and with optimum simplicity. The currently dominant paradigm cannot do so: nonlocality is an exceptional phenomenon admitted but not fully understood on the quantum level, and it is an anomaly above that level. Science requires a paradigm that recognizes nonlocal interaction as a basic feature of the universe throughout the scale of size and complexity.

An acceptable paradigm is one that provides the basic assumptions in light of which nonlocal interaction, i.e., space– and time–transcending interconnection among phenomena, emerges as an elemental fact in nature. The most plausible concept this paradigm may have to draw upon is the idea of a field: a nonlocal interaction generating or enabling field.

The concept of fields

Fields are like fishing nets so fine that their strands, though real, are not visible. When a number of fish are caught in that net, tugging at any one creates a corresponding movement in all the others.

In science the fishing net is a basic metaphor for a field. Fields are not visible in themselves, but produce visible effect. They connect phenomena. Local fields connect phenomena within a particular region of space and time, but there are also universal fields that connect phenomena through space, and possibly also time. Things interact through fields, and they may interact universally as well as nonlocally. An interaction is considered nonlocal when it transcends the known limits of effect–propagation in space and time. Nonlocal interaction is instant or quasi–instant over any finite distance.

The field that could account for nonlocal interaction in nature is not one of the known fields: it produces an effect not explained by known fields. The nature of the nonlocal interaction generating or enabling field can be discovered through the hypothetico–deductive method in science, but before attempting to discover it we review the major varieties of known fields in order to situate the additional field within their context.

Classical fields — In modern science the first fields to be postulated were to account for attraction among distant objects. Action at a distance was not acceptable—even Einstein was not happy with the idea of events occurring at a distance without some form of connection between them: he called it "spooky." If, for example, things at different points in space attract each other, classical physics sought to explain what it is that conveys the attraction between them.

In the eighteenth century physicists began to interpret the phenomenon of gravitation as the effect of a gravitational field. Initially this field was assumed to be built by mass–points in space and to act on each mass–point at its specific spatial location. Later the field concept was extended to other phenomena and was generalized. In 1849 Michael Faraday replaced direct action among electric charges and currents with electric and magnetic fields produced by all charges and currents at a given time. In 1864 James Clerk Maxwell went further: he proposed the electromagnetic theory of light. Here the EM field is universal: it accounts for electric and magnetic phenomena wherever they occur. The observed phenomena are referred to waves propagating with finite velocity in a universal field.

By the dawn of the 20th century physics had acquired four universal fields, two of them are long–range and two are short–range. These are the gravitational and the electromagnetic fields, and the strong and the weak nuclear fields. Since the middle of the century these classical fields have been joined by a variety of nonclassical fields postulated in quantum field theory.

Quantum fields — The fields of quantum physics are a complex variety: they describe entities in space and time as well as spacetime itself. The fundamental forces are described as Yang Mills fields, replacing the classical electromagnetic field. Quanta, the basic entities of the physical universe, are described by fermionic fields, and the elusive particles that endow quanta with mass are approximated as the Higgs field.

The structure of spacetime itself is generated by fields. Space is no longer an independent factor in the universe; nor is it an independent variable in field equations. As described in string theory, the structure of space is directly dependent on the conditions that define the presence of matter. Matter is not an independent reality either: it is a vibration—a "string"—in space.

According to quantum theory, at very small scales space is not smooth; it is not flat even in the absence of matter but constitutes a turbulent "quantum foam." There are mathematically cumbersome infinities associated with the quantum foam, and string theory was developed to resolve this problem. The theory eliminates the infinities by "smearing" the short–distance properties of space, smoothing the quantum turbulence. In this context the elementary entities of the universe are vibrating filaments; they appear as particles because instrumentation cannot penetrate to the required scale. (Current technology allows measurements to 10^{-18} m, whereas the Planck scale required for the string phenomenon to appear is 10^{-35} m.) If this is the case, particles are an epiphenomenon created by the limitations of our system of observation.

Strings replace the corpuscles that in general relativity curve the four–dimensional matrix of spacetime. Electrons, muons, and quarks, as well as the entire class of bosons (light and force particles) and fermions (matter particles) are not corpuscular particles but vibrational modalities defined in accordance with the geometry of spacetime.

In the sophisticated forms of string theory spacetime is "stringy": the relative points of space are themselves superstrings. Empty space is a low vibrational pattern, a "hole" in Calabi–Yau space, and the phenomena classically seen as particles appear in the intersection of the boundaries of Calabi–Yau space–holes.

The Scalar Holofield

We now turn to the field that would account for nonlocal interaction in nature.

We assume that like other fields—classical as well as quantum—the nonlocality–generating field has specific properties. Following the hypothetico–deductive method of fundamental research in science, these properties can first be "invented," but they must then be tested against observations. Does a field with the given "invented" properties provide the simplest consistent explanation of the relevant observations? In this case, does the envisaged field give the simplest consistent explanation of the nonlocality observed in the various domains of nature?

The field that satisfy these requirements has the following properties:

universality;
non–vectorial effectiveness;
holographic in–formation storage;
supraluminal effect–propagation; and
effect–production through phase–conjugate resonance.

We hypothesize that the nonlocal interaction creating field:

—is present and active at all points in space and time;

—produces effect through non–vectorial in–formation. The effect–producing agency of the field is what David Bohm called "in–formation": a physical effect–transmission that "in–forms" the recipient without conveying conventional energy and creating entropy; (i)

—conserves information in the holographic mode. Information—in the specific form of "in–formation"—is carried in the field in a distributed form: the totality of in–formation is present at all its points. The field is a natural hologram, more exactly, a natural field functioning in the holographic mode; (ii)

—transmits effects at supraluminal velocities. Observations show that nonlocal connection is instant or nearly instant at all finite distances. This suggests that the field transmits effect through longitudinal scalar waves, rather than transverse electromagnetic waves; (iii) and

—transmits its supraluminal effects through phase–conjugate resonance. For the field to produce the observed effects, it must interact with systems in space and time. The most probable modality of this interaction is phase–conjugate resonance: the scalar waves emitted by the field conjugate with the waves radiated by quanta and quanta–based systems. The resulting resonance transfers information between the field and the systems with which it interacts. (iv)

The interaction of the here envisaged field with systems in space and time produces nonlocal connection among the systems, and among the parts of systems. The effect of the field on the systems, and on the parts of the systems, can be reconstructed as follows.

The scalar waves of the field interfere with the waves emanating from quanta and quanta–based systems, and the conjugation of the respective waves creates resonance between the field and the quanta and systems of quanta with which it interacts. Since the field is universal and it transmits in–formation in the distributed mode of holograms, resonance with waves emanating from it creates connection between all quanta and systems of quanta in space and time. And given that the waves of the field are scalar, they propagate at supraluminal velocities. Therefore the connection created by interaction with the field is instantaneous over all finite distances. (v)

We conclude that a field with the properties of universality; non–vectorial effectiveness; holographic in–formation storage; supraluminal effect–propagation; and effect–production through phase–conjugate resonance responds to the requirements on a nonlocal interaction–generating field. It is a plausible hypothesis for understanding the physical processes responsible for universal nonlocal interaction.

Notes

(i) "In–formation" is not the same as information: it is not knowledge received about some fact or event, nor is it a pattern imposed on a transmission channel. Information is conveyed by in–formation, but in–formation is a different concept. It is a subtle, quasi–instant, non–evanescent and non–energetic connection between things and events at different locations in space and different points in time. It links quanta, and quanta–based systems (atoms, molecules, organisms, ecologies, stellar systems, galaxies) regardless of the space and time separating them.

(ii) Holograms code the information that produces the holographic image in a distributed form: all the information is present at all points in the holographic film or medium. Consequently the complete image appears when any small part of the film or medium is illuminated, although its definition is proportional to the size of the illuminated area.

(iii) Scalar waves are longitudinal waves like sound waves, and unlike electromagnetic waves that are transversal. The propagation velocity of longitudinal waves is proportional to the density of the medium in which they propagate (sound waves, for example, propagate faster in a dense medium such as water than in a thin medium such as air). Given the high energy density of cosmic space—the "quantum vacuum"—the propagation velocity of scalars is predictably supraluminal. The interference pattern of scalar waves stores information in the non–vectorial form of "in–formation," unlike electromagnetic waves that store ordinary information. Consequently a field of scalar waves conveys "in–formation" rather than "information," and the in–formation it conveys propagates at supraluminal velocity.

(iv) Resonance is defined in science as the harmonization of waves or vibrations: the classical forms of resonance are mechanical, acoustic, and electromagnetic. Presently also nuclear–magnetic, electron–spin, and quantum forms of resonance are recognized. The conjugation of the waves emanating from the scalar field with waves emitted by quanta and systems of quanta is a form of resonance: it transfers in–formation between the field and the systems.

(v) The conjugating waves propagate in the universal substrate still known as "quantum vacuum" although physicists recognize that it is filled with staggering amounts of energy. A number of avant–garde scientists have put forward theories that describe the processes that take place in that universal "plenum;" the pioneering research can be credited to the Russian physicists G. I. Shipov, A. E. Akimov and co–workers.[26] According to their theory vortices in the vacuum link things in space at a group–speed of 10^9 *c*—one billion times the speed of light.

The studies presented in the Annex are recent hypotheses of the origins of nonlocal interaction in dynamic processes unfolding below the level of quanta, in a domain variously identified as the Universal Quantum Field, the re–conceptualized ether, or Aether.

26 Akimov, A.E. and G.I. Shipov, 1997. "Torsion fields and their experimental manifestations." *Journal of New Energy* 2:2; and V. Ya. Tarasenko, 1992. "Models of polarized states of the physical vacuum and torsion fields." *Soviet Physics Journal* 35:3.

Chapter Five

The Akasha

If a field is responsible for nonlocal interaction between phenomena, we must form an idea of the world in which that field is manifest. However, this task is yet to be fulfilled in relation to the known fields as well. In the contemporary physical sciences fields are ad hoc postulates. They account for phenomena in specific areas of investigation, but do not disclose how the various fields are related to each other, and to the world in which they appear.

The question of "what are they" in regard to the fields and forces as well as the basic entities of nature is not adequately answered in science. Throughout the 20th century quantum physicists have dealt with the "how" of specific phenomena and left the "what" to philosophers. With few exceptions, physicists have been reluctant to consider even quanta in this context; they contented themselves with explaining their effects and processes. According to Eugene Wigner quantum physicists should be content with correlating *observations* and not concern themselves with *observables.* This proved to be a useful strategy: it enabled exploratory work to proceed pragmatically, without the burden of requiring the theories to disclose the nature of the world to which they refer. Niels Bohr advised that quantum physicists should do away with philosophy altogether: they should put a sign on their door "work in progress, philosophers keep out."

But the pragmatic strategy has led to some vexing paradoxes. In string theory, for example, correlating the observations called for assuming more than four dimensions in spacetime: ten or eleven dimensions are required by the mathematics. String theorists assume that there were this many dimensions at the origins of the universe, and that in the period of inflation following the explosion that created the universe all but four had been "choked off." But it turned out to be difficult, if not impossible, to "compactify" the extra dimensions without collapsing the remaining four as well.

The paradoxes faced by string theory are not limited to the problem of dimensions: they also concern the reality of the strings themselves. Except in the Copenhagen school of quantum mechanics, scientists assume that their theories refer to a world that exists independently of their theories. This world must be realistically conceivable, even if it is beyond the scope of the classical reality–conceptions. String theory has difficulty in producing a concept that is realistic by any standard. Strings and superstrings are vibrations, similar to musical notes. However, musical notes are produced by vibrating strings, pipes, or sounding boards, and these are physically real. String theory's strings, by contrast, float in geometrical spacetime in an unsubstantial manner reminiscent of the grin of the Cheshire Cat. We have the grin of the Cat (the vibration), while its body—the medium that would vibrate and create the various vibrational modes—is not grasped by the theory. Theory tells us that the vibrational modalities are generated by the geometrical fabric of spacetime, since it is spacetime that undergoes pinching and tearing as Calaby–Yau hole intersections define black–holes, wormholes, and elementary particles. Yet spacetime is a geometrical construct, and it is not conceivable that such a construct would vibrate, and by its vibration produce physical effect in nature.

An analogous paradox emerges in regard to the spacetime concept of relativity theory. In the special theory light is conceived as the propagation of streams of photons (or alternatively, as the deformation of the spacetime matrix), but the question, "propagation in or through what," or "deformation of what," leads to a paradox. In a conceivably realist context we would assume that if we have streams or waves traveling from one point in space to another, there is something in nature that extends between these points and carries the streaming entities or waves. The general theory of relativity comes close to answering this query, for it postulates a four–dimensional matrix that carries signals across space. In that matrix gravitation is the principal factor; it is dynamically analogous to Newton's concept of space. However, general relativity's gravitational field is background independent. Having discarded the concept of the space–filling ether, relativity theory is properly called a theory of *relativity*. Its spacetime concept is not grounded in a realistically conceivable universe. Given Einstein's insistence that his theory refers to the real world, this is paradoxical.

Stephen Hawking's conversion from the basic realism that held that the existence of the universe is independent of our theories of it, to a "model–dependent realism" that denies an observed–independent theory and a theory–independent reality, reflects the prevailing attitude in the physics community.[27] The problem is fundamental. The "Standard Model"—relativistic Quantum Field Theory (QFT) transformed into Quantum Electrodynamics (QED), which led in turn to the Electroweak Theory that seeks to unify the electromagnetic force with the weak force of nuclear interaction. The force of gravity could not so far be satisfactorily quantized, and quantum gravity could not be unified with Quantum Field Theory. The gap remains unbridged between the supersmall entangled world of the quantum and the relativistic world that reigns at very large dimensions.

A new paradigm in physics would need to ground the theories of physics in an independently existing and basically realistic (that is, humanly conceivable) universe. This requirement applies to all the fields of relativity theory as well as of quantum field theory. It also applies to the scalar holofield.

The concept of a hidden dimension

Responding to the requirement for an observation– and theory–independent universe calls for reviving an ancient and quasi–perennial intuition: the existence of a hidden dimension in the universe. This "hidden" plane is the locus of the fields and forces of nature, together with the laws and relations that tie them into a coherent whole. Fields, as already remarked, are not observable in themselves, but produce observable effect. The domain of fields is not observable; we know it only through its effect on the observable domain.

27 Stephen Hawking and Leonard Mlodinow, *The Grand Design*. New York: Bantam, 2010.

An intrinsically unobservable domain as the deeper or higher reality in the universe is a perennial intuition in the history of science and philosophy. Philosophers of the so–called mystical branch in Greek metaphysics (the Idealists and the Eleatic school, comprising thinkers such as Pythagoras, Plato, Parmenides, and Plotinus) differed on many points, but were united in the affirmation that there is a deeper or higher dimension in the world. For Pythagoras this was the *Kosmos,* a transphysical, unbroken wholeness, the prior ground in which matter and mind, and all being in the world arises. For Plato this was the realm of Ideas and Forms, and for Plotinus "the One." The perceived world is illusory, ephemeral and short–lived, while the fundamental dimension is eternal, and eternally unchanging.

At the dawn of the modern age Giordano Bruno, the pioneer who challenged the Aristotelian–medieval world picture and then paid for this "heresy" with his life, brought the concept of the hidden dimension into the ambit of science. The infinite universe, he said, is filled with an unseen substance called *aether* or *spiritus.* The heavenly bodies are not fixed points on the crystal spheres Aristotelian and Ptolemaic cosmology, but move without resistance through this unseen cosmic substance under their own impetus.

In the 19th century the French physicist Jacques Fresnel affirmed the idea of a space–filling and in–itself unobservable medium called "ether." In Fresnel's theory the ether is a quasi–material substance in which the movement of heavenly bodies produces friction. If the Earth moves through the ether, the light that reaches it from the Sun should display the effects of "ether–drag": toward the light source the beams should be reaching the Earth faster than in the opposite direction.

At the turn of the 20th century Michelsen and Morley measured the speed of light in the direction of the rotation of the Earth and also in the contrary direction but failed to discover a difference. The physics community took this as evidence that the ether does not exist, notwithstanding Michelsen's warning that the experiments disproved only a particular mechanistic theory of the ether and not the concept of a space–filling medium that would transport light as well as other fields and forces.

When Einstein published his special theory of relativity, the theory of the ether was discarded: it was no longer necessary. All movement in space (that is, in the four–dimensional spacetime continuum) was said to be relative to the given reference frame; it is not movement against a fixed background.

However, the ether, as an unobservable plane of reality underlying observable phenomena, came back to physics through the back door. Theoretical physicists began to trace the fields and forces of nature to common origins in a unified, and later in a grand–unified and then super–grand–unified field. An in–itself unobservable ground that produces observable phenomena entered the assumptions that underlie the theories. David Bohm's theory of the "hidden variables" was an explicit recognition of this dimension: his "implicate order" is an unobservable holofield that grounds the quantum–level events we encounter in the explicate order.

The Akasha

A hidden dimension in the universe, a basic feature of traditional cosmologies, has resurfaced at the leading edge of contemporary science. An adequate paradigm recognizes this dimension and states the basic principle by which we can explain it.

In writings that span more than three decades, this writer suggested that the factor in nature responsible for the observed nonlocality is a particular field. He first called the nonlocality generating field *psi field* (since it accounts among other things for a variety of so–called "psi"–phenomena), and then named it *quantum/vacuum–interaction* (QVI) field (given that through this field quanta interact with the "vacuum" that embeds them). In 2004 he renamed the field *Akashic field*, in recognition of the age–old intuition of an Akashic record. He now realizes that the element in nature that accounts for nonlocal interaction is more than a particular field: it is the hidden domain that harbors all fields. The name "Akasha" applies to this domain, consistently with the original insight.

The ancient *rishis* held that the Akasha is the fundamental element of the cosmos, the ground of the four elements *vata* (air), *agni* (fire), *ap* (water), as well as *prithivi* (earth). It holds these elements within itself, but it is also outside of them, for it is beyond space and time. According to Paramhansa Yogananda, the Akasha is "the subtle 'background' against which everything in the material universe becomes perceptible."

In his classic *Raja Yoga,* Swami Vivekananda gave the following account of the Akasha:

> It is the omnipresent, all–penetrating existence. Everything that has form, everything that is the result of combination, is evolved out of this Akasha. It is the Akasha that becomes the air, that becomes the liquids, that becomes the solids; it is the Akasha that becomes the sun, the earth, the moon, the stars, the comets; it is the Akasha that becomes the human body, the animal body, the plants, every form that we see, everything that can be sensed, everything that exists. It cannot be perceived; it is so subtle that it is beyond all ordinary perception; it can only be seen when it has become gross, has taken form. At the beginning of creation there is only this Akasha. At the end of the cycle the solid, the liquids, and the gases all melt into the Akasha again, and the next creation similarly proceeds out of this Akasha.[28]

The Akasha is the all–encompassing element that both underlies all things and becomes all things. In its subtle aspect it *underlies* all things, and in its gross aspect it *becomes* all things. In its subtle aspect the Akasha cannot be perceived; it can only be observed in its gross aspect, where it has become the things that emerge and evolve in space and time.

In the context of science, the Akasha is the hidden dimension of the universe. It is the dimension out of which the observable universe arose at the beginning of a cosmic cycle, and into which it will again re–descend. This is the basic concept of the scheme that will enable scientists to live up to Einstein's ideal of science: to tie together the observed facts—all the observed facts, including those that are presently anomalous—with optimum consistency and simplicity.

---

28 Vivekananda, Swami. *Raja Yoga*. Calcutta: Advaita Ashrama, 1982.

PART III

THE AKASHIC VIEW OF THE WORLD

A new paradigm is needed to account for the coherence and nonlocality discovered in the various fields of the empirical sciences. The paradigm that does so recognizes a hidden dimension in an independently existing and humanly conceivable universe. This paradigm revives and reconceptualizes a basic insight: Akasha, as the fundamental element of the cosmos.

In regard to the worldview that dominates today's mainstream science, this concept is revolutionary. The five chapters of Part Three explore and articulate this "worldview (r)evolution."

Chapter Six outlines the emerging concept of the cosmos: an Akashic cosmology.

Chapters Seven describes the new concept of cognition: the corresponding (r)evolution in the theory of knowledge.

Chapter Eight assesses the basic nature of our freedom to determine our life and actions in the world, offering a humanly relevant ontology.

Chapter Nine outlines a science-based grounding for our values and aspirations: the foundations of a new ethics.

And Chapter Ten advances a (r)evolutionary concept of mind: the Akashic theory of consciousness—of individual consciousness and also of cosmic consciousness.

Chapter Six

Cosmos

Theorizing about the coherent whole of reality the Greeks called "cosmos" is an integral part of our intellectual heritage. In the past it was based on personal insight and individual experience. Since the dawn of the modern age it has been attempted also on the basis of science.

Cosmology in the sense used in the history of thought—not as a physics specialty but as an account of our most basic concept of reality—is particularly sensitive to the choice of its paradigm. When that paradigm is mechanistic, the resulting cosmology portrays the world as a giant mechanism. When the paradigm is vitalistic, the picture that emerges is of the world as a cosmic organism. And if it is idealistic or spiritualistic, reality is perceived as the manifestation of a cosmic mind or consciousness.

For the past several hundred years, Western science was dominated by the materialist Newtonian paradigm. Cosmologies based on this paradigm envisaged the world as a vast mechanism, running on the energy—the negative entropy— with which it was endowed at its birth. The world is running inexorably down, toward the state of maximum entropy where no irreversible processes are any longer possible and hence nothing new can come about.

The staggering quantum sea discovered at the subquantum level of the universe challenged the Newtonian concept of the closed clockwork universe. Our cosmology has changed. We can now envisage the universe in which we live not as a single, one–way process from Big Bang to Big Crunch but as a cyclic process with phase transitions: Big Bounces. The cycles produce evolution followed by devolution in each cycle. But the information accumulated in each cycle does not vanish: it "in–forms" the successive cycles. Information is the software that governs the cycles; it is the software of the cosmos. It is an unobservable, "hidden" dimension in the cycles. This is the basic concept of Akashic cosmology.

First principles

Akashic cosmology views the universe as an integral system evolving in the interaction of two dimensions: a hidden dimension, and an observable dimension., The hidden dimension is the Akasha: we refer to it as the A–dimension. The observable dimension is the manifest or M–dimension.

The two dimensions are related to each other as the body of water that makes up the sea is related to the waves that appear on its surface. The waves are generated by the body of water that underlies them; they are not separate entities. But on first sight they could be mistaken for being that: we see only waves interfering with waves and creating interference patterns. It takes a deeper understanding to realize that the waves are not a separate reality; they are patterns produced by the sea. The waves we see are not only waves *in* the sea; they are waves *of* the sea.

The relation between the water of the sea and the waves that appear on its surface is a simplified metaphor for the relation between the A– and the M–dimensions. The underlying body of water—the A–dimension—produces the waves that appear on its surface, in the M–dimension. The waves and the body of water in which they appear are in interaction. The body of water produces the waves, and the waves shape the body of water. In this cosmology we say that the A–dimension "in–forms" the M–dimension, and the in–formed M–dimension "de–forms" the A–dimension—modifies its in–forming potential.

The particles and systems of particles in the M–dimension are a manifestation of the A–dimension, and they are affected by both dimensions. Every particle and every system has what Alfred North Whitehead called a "physical pole" through which it is affected by other particles and systems of particles, and a "mental pole" by which it is affected by the A–dimension. Whitehead called these affects "prehensions"—the effect of the rest of the world on every element of the world.

Particles and systems in the M–dimension prehend the rest of the world in two modes. They prehend the M–dimension through the fields and forces and the corresponding laws of nature that govern existence in the manifest world, and they prehend the A–dimension through the subtle effects Plato ascribed to the realm of Forms and Ideas, Whitehead to eternal objects, and Bohm to the implicate order. The former are the recognized effects of the external world on physical bodies and the sensory organs that may be associated with those bodies, and the latter are the spontaneous instincts and intuitions that informed traditional cultures but are mostly suppressed or ignored in the modern world. They are conveyed by quantum–level interactions and were not recognized as effective agencies until recently.

The M–dimension and the A–dimensions are diachronically as well as synchronically related. Diachronically, the A–dimension is prior: it is the generative ground of the particles and systems of particles that emerge in the M–dimension. Synchronically, the generated matter–energy particles and systems of particles are linked with the A–dimension through a constant bi–directional interaction. In one direction, the A–dimension "in–forms" the particles and systems that populate the M–dimension. In the other direction the in–formed particles and systems "de–form" the A–dimension. The latter is not an eternal and immutable domain of Platonic Forms, but a dynamic matrix progressively structured by interaction with the M–dimension.

Particles and systems of particles in the M–dimension are not discrete entities, disjoined either from each other or from the A–dimension. The things that populate the M–dimension are soliton–like waves, nodes, or crystallizations of the A–dimension. It is in the A–dimension that they emerge, it is with the A–dimension that they "dance" and evolve. And it is into the A–dimension—the Akasha—into which they re–descend when the current cycle of the metaverse comes to an end.

This two–dimensional interactive process was anticipated in Hindu cosmology. There the almost infinitely varied things and forms of the manifest world are reflections of oneness at the deeper, non–manifest level. At that level the forms of existing things dissolve into formlessness, living organisms exist in a state of pure potentiality, and dynamic functions condense into static stillness. All attributes of the manifest world merge into a state beyond attributes. Time, space, and causality are transcended in a state of pure being: the state of Brahman.

Brahman, though undifferentiated, is dynamic and creative. From its ultimate "being" comes the temporary "becoming" of the things of the manifest world, with their attributes, functions, and relationships. The *samsara* of being–to–becoming, and then of becoming–to–being, is the *lila* of Brahman: its play of creation and dissolution. The absolute reality of Brahman and the derived reality of the manifest world form an interconnected whole: together they are the *advaitavãda* of the cosmos. Brahman is the fundamental reality. The things that appear in the manifest world have secondary reality—mistaking them for real is the illusion of *maya*.

In Akashic cosmology the Akasha takes the place of Brahman. The Akasha, though "hidden" in regard to observation, is the ultimate reality; it harbors all of nature's fields and forces. The effect of the fields and forces on particles, atoms, organisms and galaxies are the ceaseless "play" of the universe on the manifest plane.

Evolution *in* the cosmos and the evolution *of* the cosmos

Astronomical and astrophysical evidence leave no doubt that our universe is an evolving and not a steady–state system. Its matter–content was created in the burst that marked the transition from the previous to the current cycle of the metaverse, and it will vanish in the transition to the next cycle.

Across the metaverse's cyclic universe–creating process, bi–directional interaction between the A– and the M–dimensions constitutes a continuous loop of action and reaction. This loop creates increasing coherence in the M–dimension, and accumulates in–forming potential in the A–dimension.

Systems of particles, and systems of systems of particles, evolve in each cycle of the metaverse, but their evolution is limited by physical conditions in the given cycle. The cycles are finite, and physical conditions in them are not indefinitely conducive to the evolution and persistence of complex systems. Thermal and chemical conditions are suitable for the build–up of such systems only during the expansionary phase of the cycles. As expansion reaches an apex and gives way to contraction, physical conditions for the evolution and persistence of complex systems become unfavorable. During the supercompacted final phases of a cosmic cycle only the stripped nuclei of atoms persist; and then also those die back into the "quantum vacuum," the enduring Akashic matrix of the cosmos.

Current cosmological models (in particular the quantum loop gravity cosmology reviewed in Chapter 3) suggest that the structure of the quantum vacuum is conserved from cycle to cycle. "Vacuum" does not mean empty space here, or even the turbulent energy foam postulated in quantum field physics; it refers to the cosmic matrix: the A–dimension. The structure of that matrix is not cancelled in the transition between cycles but conserves the in–forming potential it has accumulated in the foregoing cycles.

The interaction of evolving/devolving systems in the M–dimension with the enduring matrix of the A–dimension introduces direction into the trans–cyclic process. In each cycle the cosmic matrix in–forms the systems of the M–dimension, and the evolving systems de–form the matrix. Because the in–forming potential of the matrix is conserved across the cycles, it accumulates across the cycles. Thus the evolutionary process is cyclic for the M–dimension, and cumulative as regards the A–dimension.

A learning curve unfolds across the cycles. As the in–forming potential of the A–dimension accumulates, systems in the M–dimension become more and more in–formed. They evolve to greater and greater complexity and coherence, reaching equal levels of evolution in shorter times, or higher levels in equal times. In biological systems this evolution is manifested as increasing coherence within and among organisms, while on the astronomical level it comes to light as the increasing coherence of galactic macrostructures and evolutionary processes.

Particles and systems of particles evolve and devolve in each cycle and build sequentially to higher and higher levels of complexity before breaking down and vanishing, and the cosmic matrix persists across the cycles. It is constantly de–formed by interaction with the systems, and as it conserves its deformation, its potential to in–form particles and systems is growing.

If evolution within and across the cycles is irreversible, the sequence of cycles in the metaverse cannot be infinite. It ultimately reaches a final Omega cycle, where evolution within and among the cycles reaches a zenith. At the end of that cycle systems in the manifest dimension achieve the highest state of complexity and coherence, and the A–dimension attains the highest potential for in–forming systems in the manifest dimension.

Chapter Seven

Cognition

We live in a dynamic, nonlocally interconnected universe and are in constant interaction with the rest of this universe. These interactions provide us with the free energies and free–energy containing material resources we need to live and to thrive. The signals we receive from the rest of the universe orient us in regard to accessing the energies and resources we need. We now ask, how do we receive and assess the signals that reach us, and what is the nature and the scope of these signals?

Framing these questions in reference to the Akasha paradigm brings to light new facts and facets: a (r)evolutionary theory of human cognition. Here we review the principal features of this theory.

The Akashic Theory of Cognition

We are systems in the manifest dimension of the cosmos, and we interact both with this dimension and with the underlying Akashic dimension. The information we receive from the manifest dimension is in the form of wave propagations in the electromagnetic spectrum and in the air; while the information we receive from the A–dimension takes the form of quantum wave propagations. The former are conveyed by the bodily senses, and the latter are decoded by subneuronal structures in our brain.

Everyday experience is dominated by information conveyed by our senses: the sights, sounds, smells, flavors, and textures of the world that surrounds us. Until recently most scientists maintained this is the only information we receive from the world beyond our body. The motto of classical empiricism was "everything that is in the mind was first in the eye." This reduced human experience to sensory data. New developments in cutting–edge neuroscience show that the classical concept is outdated.

Sensory information is processed by neurons connected by synapses in a neuroaxonal network. This network was considered the sole basis of our perception of the world. It turns out, however, that the neuroaxonal network is not the only mechanism that processes signals from the world. There is a vast hierarchy of networks below the level of this network, extending all the way to quantum dimensions. This hierarchy of networks likewise processes signals from the world, albeit in a different mode.

The brain's subneuronal networks are built of cytoskeletal proteins organized into microtubules. The microtubular networks are connected to each other structurally by protein–links and functionally by gap junctions. Operating in the nanometer range, the number of elements in these subneuronal networks substantially exceeds the number of elements in the neuroaxonal network: there are approximately 10^{18} microtubules in the brain, compared with "merely" 10^{11} neurons.

Neurophysiologist Stuart Hameroff and physicist Roger Penrose suggested that the networks of microtubules process information on the quantum level.[29] They realized, however, that microtubules may be too coarse–grained for this function: it is more likely that it is the periodic lattice within the network of microtubules that is responsible for quantum–level information processing. This "microtrabecular lattice," a part of the cytoskeletal lipoprotein membrane, is a web of microfilaments seven to nine nanometer in diameter. Researchers Ede Frecska and Eduardo Luna maintain that this is the cerebral structure that processes quantum–level signals reaching us from the environment.[30]

There are two structures in the brain that process the signals we receive from the world: the macro–level neuroaxonal networks,

29 Stuart Hameroff, 1987. *Ultimate Computing*, North Holland Publishers; and Roger Penrose, 1996. *Shadows of the Mind: A Search for the Missing Science of Consciousness*. Oxford University Press. Also see Stuart Hameroff, Roger Penrose, et al. *Cosmology, Consciousness and the Universe: Quantum Physics, Evolution, Brain & Mind*. 2011. Cosmology Science Publishers.

30 Frecska and Luna, op. cit.

and the subneuronal quantum–level network–hierarchy. Thus we have two modes of experiencing the world available to us and not one: the mode Luna and Frecska call "perceptual–cognitive–symbolic," and the mode they call "direct–intuitive–nonlocal." The former processes signals originating in the manifest dimension, and the latter processes signals that originate in the A–dimension. The A–dimension also in–forms our brain. This in–formation represents an additional and hitherto largely ignored effect of the rest of the world on our brain and body.

The A–dimension in–forms us by nonlocal interaction with the subneural structure of our brain. These interactions are in the form of the interference of scalar waves originating in the A–dimension with quantum waves produced by the brain. When the interfering waves enter into resonance, the conjugation of the wave–patterns transfers information between the scalar holofield of the A–dimension and our brain. This information is nonlocal: it could have originated anywhere in the universe, at any time. It is processed by quantum–level arrays in our brain without passing through our senses.

Sense–conveyed information gives rise to the sights, sounds, textures, and odors that dominate our everyday consciousness, and quantum–resonance based information conveys subtle effects such as hunches, images, insights, and intuitions. These do not always reach our conscious awareness.

Perception is a highly selective process. The brain is a collection of nerve cells that function as multi–layered frequency receptors, and these are selective as regards the signals to which they respond. Due to conditioning from early on in life, each receptor becomes wired to receive a particular frequency. The act of "tuning in" to the information coming to us from the world means picking–out the frequency patterns that are familiar from an ocean of patterns and frequencies that is unfamiliar—and thus unrecognized and ignored.

As the receptors tune–in to a particular set of frequencies, a pattern–recognition response is generated: the information–processing networks of the brain interpret the given pattern in accordance with the interpretation that has been established for it. By tuning into the same pattern time and time again, a particular interpretation is reinforced. Thus our sensory perception of the world is created moment by moment as our brain scans the available frequency bands and produces the interpretations associated with the familiar patterns. Selectivity based on repeated patterns is typical for all elements and aspects of our experience: we have difficulty in recognizing, even perceiving, unfamiliar patterns.

This kind of selectivity operates also in regard to the signals processed by the subneuronal networks of our brain. Selectivity here concerns the transfer of signals from the quantum–level receiving networks of the brain to the cognitive centers responsible for our waking consciousness. These signals are mostly unfamiliar and their transfer to conscious awareness is effectively blocked. But they in–form the subneuronal networks of the brain and their transfer to the cognitive centers can be facilitated. For modern people (unlike for most traditional cultures), this requires entering an altered state of consciousness.

In an altered state quantum–resonance based information can reach the level of conscious awareness. This results in subtle intuitions, such as those conveyed occasionally in dreams, white–dreams and daydreams, and in the images, sounds and messages that emerge in deep prayer and meditation. The hypnagogic states that mark the transition between sleep and wakefulness are particularly apt to convey quantum–resonance based information. The transfer of such information to waking consciousness can be purposively fostered, among other things by cultivating meditation and prayer, calm introspection, communion with nature, and esthetic experiences.

Altered states are often catalyzed by unusual, life–transforming experiences. Astronauts who saw our planet from space reported experiences of a distinctly nonlocal nature. Edgar Mitchell, captain of the lunar module of Apollo XIV, was the second man to walk on the moon. While he was in space he had an epiphany that changed his life. He beheld the network of life on Earth as an interconnected whole. On returning to Earth he founded the Institute of Noetic Sciences, dedicated to research on nonlocal experiences and devoted the rest of his life to understanding these experiences. He wrote, the answer to how spontaneous events lead to deep insights with behavioral modifications is the Akashic field, where it is augmented continuously by quantum holographic information.[31] The intuitions that surface in these experiences should be considered our first rather than our sixth sense, because this sense was developed in nature long before humans began to ascribe the information they receive exclusively to their senses.

Entering altered states of consciousness, whether through unusual experiences or through prayer, meditation, esthetic experience or the contemplation of nature, synchronizes the left and the right cerebral hemispheres. Experiments show that in altered states the electrical activity of the hemispheres becomes harmonized: the patterns that appear in one are matched by patterns in the other. This contrasts with ordinary states of consciousness where the hemispheres function nearly independently of each other. Moreover when people enter deeply meditative states together not only their own hemispheres become synchronized; synchronization extends to the entire group. In experiments by Italian brain researcher Nitamo Montecucco, eleven of a group of twelve meditators achieved a level of cross–hemispheric as well as transpersonal synchronization that exceeded 90 percent. Yet the meditators sat with closed eyes, in silence, and did not see, hear, or otherwise perceive each other.[32]

Synchronization comes about even when the meditators are on distant locations. With increasing distance the level of cross–hemispheric and transpersonal synchronization falls off, but it remains considerably above random probability.

The remarkable experiences that surface in deeply altered states of consciousness have been described by many psychiatrists and psychotherapists. The experiences include spontaneous contact with persons, things and events we could not have experienced in the classical sensory mode. Psychiatrists practicing regression therapy report that some of their patients access people and events from what appears to be a previous life. So–called "psi–phenomena" occur more frequently in altered than in normal states; telepathic and clairvoyant abilities are surfacing.

---

31 Mitchell, Edgar, 1977. *Psychic Exploration. A Challenge for Science*. New York: G.P. Putnam.

32 Montecucco, Nitamo, 2000. *Cyber: La Visione Olistica*. Rome, Italy: Mediterranee.

Altered–state experiences are distinctly nonlocal. Psychiatrist Stanislav Grof found that in these states people experience a loosening and melting of the boundaries of their individual ego and a sense of merging with other people and other forms of life. In deeply altered states many people report an expansion of their consciousness to the extent that it encompasses all life on the planet.[33]

* * * * *

The classical concept of cognition needs to be revised. We are not limited to seeing the world through "five slits in the tower" — we can open the roof to the sky. We transcend the range of our sensory organs when we become conscious of the signals processed by our brain's subneuronal arrays. In that mode our experience includes nonlocal elements. To transfer the information received in this mode to conscious awareness most of us in the modern world need to enter altered states of consciousness. In such states the two hemispheres of our brain function coherently, and information that reaches the brain's quantum–level arrays finds its way to the centers that code our waking consciousness. The flow of our consciousness becomes enriched with feelings and sensations that connect us with other people and with nature, and ultimately with the whole of the universe. We connect with the A–dimension, the matrix of the cosmos.

33 Grof, Stanislav, 2000. *Psychology of the Future: Lessons from Modern Consciousness Research.* Albany: State University of New York Press.

Chapter Eight

Freedom

The Akasha paradigm implies a major (r)evolution in our view of the world. There is more to the universe than we had thought, and there is more to the information we receive from the world than we had assumed. And there is also more to our freedom in the world than the paradigm of mainstream science would have us believe.

We are an organic part of a nonlocally interconnected and interacting universe. In this world there is no absolute freedom, for such freedom would require a total lack of ties to the world. There is, however, a significant, and even a surprising degree of freedom available to us: the freedom to make use of our ties as we ourselves may wish.

Interconnection and interaction does not rob us of the freedom to determine how we are and how we act. Freedom does not reside in being "free from" external influences, but in being "free to" act the way we decide in regard to the influences that act on us.

Self–determination and not indetermination is the hallmark of freedom. A complete absence of external influences would mean autarky for complex systems and randomness for simple ones. A die is expected to be free from all influences beyond the impetus that has been imparted to it and its friction with the table, since further influences would bias its roll. But even if a die is free from such influences, it does not possess freedom, for it does not choose the way it rolls. Freedom resides not in an absence of external influences, but in the determination of the response to the influences. In that sense human beings have significant freedom in the universe. We can decide how we respond to the influences that act on us in the world.

There are factors in the determination of every living system's response that are external to it, and others that are internal. The external factors influence the way the system acts, reducing the scope of its self–determination. The internal factors concern the selection of the influences that act on the system, and the selection of the system's response. These factors are elements of freedom for the system: they enhance its level of self–determination. The relative weight of the external vs. the internal factors differentiates between the freedom of an amoeba to move in relation to its food supply, and the freedom of a human being to select the values and goals by which he or she decides to live.

Freedom in the world is neither nil, nor full. It is a matter of degree. In simple systems the element of self–determination is modest: the systems respond to external influences without exercising significant control either over their reception of the influences or their selection of the response. Theirs is a relatively direct *reaction* to the *action* of the external world, typical of physical and physical–chemical systems. In biological systems the element of self–determination becomes significant. The action of the world is received as a series of stimuli, and these can be coupled to a wide range of responses that are either supportive of the life and evolution of the systems, or are detrimental for them. In relatively simple systems the stimuli are perceived as an undifferentiated "feel" of the world, while in human beings the basic feel is differentiated as a series of articulated perceptions with conscious as well as subconscious, rational as well as emotive elements.

As already noted, we receive two kinds of signals from the rest of the world; two varieties of stimuli. One kind originates with the fields and forces of nature, including wave–propagations in the air and in the electromagnetic spectrum. These link us with the manifest world, the M–dimension. The other kind of signals are more subtle but just as real: they are the signals by which the A–dimension "in–forms" our brain and our body.

We select our response to these signals in accordance with our needs and wants, whether real or perceived. We admit some of the signals to our consciousness as bona fide perceptions of the world, and exclude others as irrelevant or unreal. In the modern world we exclude from waking consciousness the great majority of the subtle signals that reach us from the A–dimension. This reduces our capacity to orient ourselves in the world; limiting the scope of our freedom.

In the web of life on this planet we enjoy the highest level of self–determination, and hence of freedom. We can exercise considerable control over the selection of the signals to which we respond, and the way in which we respond to them. This is crucial for our life and wellbeing. As other living systems, we maintain ourselves in a dynamic state far from thermal and chemical equilibrium through the intake and processing of information, energy, and matter. This calls for constant, highly focused sensitivity and accurate response in regard to the relevant flows and resources. The right flows and resources need to be selected at the right time, and they need to be coupled with the right responses, lest free energy becomes depleted in our system and our vitality becomes impaired.

The more complex a system, the more decisive for it is how it selects its signals and how it couples it with responses. We achieve this coupling through the information processed by our brain and nervous system. Freedom is significant for us to the extent that this information is well processed: that the signals are properly selected, clearly differentiated, and accurately coupled with life–sustaining responses.

The selection of the signals that come to us from the M– and the A–dimensions is one aspect of our freedom: the aspect of self–determined selectivity in regard to the influences that act on us. Another aspect resides in the selection of our response. In comparatively simple systems, such as the more primitive forms of life, the response to external stimuli is in the form of a mainly pre–programmed chain of action and reaction. In us, however, the response is conditioned by a series of "intervening variables," and these are partially, even if not fully, under our conscious control.

There is a vast array of sub– or non–conscious factors that define our response to any signal. This array includes tacit preferences and unexamined values, cultural predispositions and a range of acquired or inherited leanings, preconceptions and prejudices. These shift the factors that determine our response to the world from the external world to us: to our values, worldviews, and preferences.

If we allow a broad range of the signals that reach us to penetrate to our consciousness, we extend the freedom we enjoy in the world: we create a wider basis for choosing our response. We adopt consciously envisaged goals and objectives, entertain consciously adopted values, and bring our goals and values to bear on the conditions that shape our life and decide our destiny.

In the final count human freedom is two–fold. It resides in our ability to select the signals that reach us from the world, and in our ability to select the way we respond to the signals. Both elements of our freedom are significant, and both can be consciously enhanced. We can respond not only to the stimuli that reach us from the M–dimension, but also to those that come to us from the A–dimension: the subtle intuitions that surface in slightly or deeply altered states of consciousness. The wider the range of the influences to which we respond, the more extensive is our freedom.

In addition to external influences, signals from the M– and the A–dimensions, we have the freedom to act in light of signals that we envisage rather than actually receive. We can shape our thinking and our behavior in accordance with our envisagement of the future and our recall of the past. We are not limited to acting and reacting in the here–and–now. Not only can we react, we can *proact.* We have the potential for the highest degree of freedom of any system in this corner of the galaxy.

Chapter Nine

The Good

We have genuine freedom in the world. The world does not determine what we do; in the final count we determine our thinking and our behavior ourselves.

We are conscious beings, and our freedom includes the freedom to consciously choose the way we think and act. If we have the freedom to choose the way we think and act, we have the responsibility to choose it wisely. Self–interest will make us act so as to realize our immediate desires and aspirations. But are these desires and aspirations in our best interests? Are they also in the best interest of the world around us? What are the truly good choices? What criteria can we apply to get a reliable answer?

These questions raise the issue of morality—the moral implications of the way we think and the way we act. The Akashic worldview offers the basis for a new ethics.

Criteria for The Good

The moral choice is to strive for value, ultimately for the highest value philosophers called The Good. But what is The Good? Are there criteria for defining it?

The criteria for deciding what is good and what is bad has been debated for millennia. No definitive consensus has emerged. In Western philosophy the view of the classical empiricists has prevailed: judgments of good and bad are subjective; they cannot be decided unequivocally. At the most they can be related to what a given person, a given culture, or a given community holds to be good. But that, too, is subjective, even if it is subjective in relation to a group: then it is *inter*subjective.

However, in the context of the Akashic worldview we find objective criteria for defining the nature of The Good. These criteria do not carry the certainty of axiomatic statements, but they are not merely subjective or intersubjective. They are as objective as any statement can be about the world. They refer to the conditions that ensure life and wellbeing in the nonlocally interconnected and interacting universe. These conditions can be briefly restated.

Living organisms are complex systems persisting in a dynamic state far from thermodynamic equilibrium. What is good for them is that which enables them to maintain themselves in this condition. Specifying all the things this implies and embraces would mean listing the entire gamut of conditions and resources a living species needs for living and for thriving. But there are basic principles that can be succinctly stated.

All living systems need to ensure a reliable access to the energy, resources, and information they require in order to maintain themselves in their milieu. This means that the systems must safeguard the coherence of their cells and organs, and the molecules that make up their cells and organs. The coherence of the whole organism is the precondition of life. For an individual organism, acting to safeguard its coherence is the right way to act. It is good for it; but it is not "The Good."

The Good, the highest value for a system that exists in close interaction with its environment, is not an individual good. It is not what enhances it's own coherence, but what enhances his or her coherence together with the coherence of other humans and other living systems in the environment. This calls for enhancing our own coherence in harmony with the coherence of the web of life on this planet.

Maintaining our coherence calls for a high level of cooperation between all the cells and organs that make up our organism. Despite the diversity of our cells and organs—indeed, *because* and *through* their diversity—and thanks to the many local as well as nonlocal connections among them, we can perform the almost miraculous feat of maintaining ourselves in life's inherently unstable state far–from–equilibrium. This feat is achieved because all our cells and organs can be effectively and precisely coordinated. Any breakdown in coordination signifies decreased viability. It is a sign of weakness in our organism, a prelude to disease.

A healthy organism is organically coherent, with all its parts tuned together to maintain the whole system in its environment. But for achieving and maintaining organic coherence, a healthy environment is *sine qua non*. The Good for an organism is organic coherence in an organically coherent environment.

To maintain our own coherence and to promote coherence in others is objectively good: it is the moral way of acting. Conversely it is objectively bad, and hence immoral, to impair our health and destroy or damage the coherence of others in our environment.

Acting destructively vis–à–vis our environment indicates a breakdown in the flow of information that connects us with the world. In our day the most common symptom of this breakdown is a sense of separation. Modern people feel separate from others around them, and from nature. This is a dangerous condition. Species that do not possess a higher nervous system cannot make this mistake: they are more directly connected with the world around them and cannot opt out of it. We, on the other hand, can fail to recognize our connections; we can entertain the illusion young people call "duality." The world and I are not one—we are radically and categorically separate. This has grave consequences. People who feel themselves separate are prone to treat others and their environment as a means to satisfy their own ends and aspirations without regard for the health and wellbeing of others and the coherence of all systems in their environment.

Another symptom of this breakdown is a disregard for what traditional and sensitive—and not just romantic—people call "love." Love is not just sexual attraction and the desire for satisfaction. Love is connection; belonging; oneness—with others, with nature, with the universe. A life without love is a disjoined and lonely life. Love is the longing for coherence between oneself and the rest of the world. It is healthy and moral both subjectively, as the expression of a healthy inclination and instinct, and objectively, as it responds to a urgent and real requirement in the world.

The breakdown of connection between oneself and the world is bad for one, and bad for the world. The self–concerned behavior inspired by it is a wrong use of the freedom of human beings in the universe.

The contrary also holds true. Love, nonduality, rationally recognizing and spontaneously seeking our connection with the world is objectively good. It is individually health–enhancing and socially and ecologically sound. It is a moral use of human freedom.

Striving to maintain our coherence in harmony with the coherence of others is the highest value in life. It is The Good—the optimum use of the freedom granted to a living system with an evolved consciousness.

Chapter Ten

Consciousness

Many philosophers and most theologians maintain that there is a mental or mind–like element in the world. Akashic cosmology views the cosmos as an integral whole evolving through the interaction of its hidden and manifest dimensions. Could this self–evolving whole have a mental aspect?

Matter and mind in the cosmos

We experience the materiality of the world in our inspection and recognition of other things around us, and we experience mind in the world by introspecting on our own experience and assuming that other people—and perhaps all living things—have a similar kind of experience.

Are mind and matter both present in the world? And are they equally fundamental? Or could it be that one is the reality and the other a mere illusion? These questions have been debated for millennia. The answer, sometimes tentative but more often dogmatic, has been usually framed in the philosophical doctrines of *materialism, idealism*, and *dualism.*

For materialism all tangible things in the world are material—they are made up of a universal substance called matter.

For idealism all things, whether tangible or not, are mental or mind–like, given that mind is the basic, and possibly the only, reality of the cosmos.

For dualism both matter and mind are real, and human beings—and possibly all living things—are material as well as mental.

All these positions call on lived experience for support.

Consider the materialist position. The world to which our experience seems to refer is a world of solid things such as atoms, and the many things composed of atoms. These are material things: "matter" in one combination or another. Mind and consciousness are epiphenomena, byproducts of the brain, a complex material thing in its own right.

Let us consider now the principle of idealism. We experience the world through our consciousness, and we cannot exclude the possibility that the world we experience is not only given in or through our consciousness, but that it *is* our consciousness. After all, our stream of consciousness consists of perceptions, volitions, feelings, and intuitions, and the world could be a construction based on these elements. This is a strong possibility: René Descartes and other famous philosophers analyzed the stream of human consciousness and failed to find convincing proof that there is a world that exists independently of it. The only thing that they could not doubt is the existence of the act of experiencing itself: *cogito ergo sum.*

Materialist and idealist philosophers affirm the one or the other of these positions to the exclusion of the other. Dualists claim that both matter and mind are fundamental elements in the world, although they do not agree whether mind is present in all things or only in human beings. In humans a material brain is connected with, or perhaps it produces, an immaterial consciousness.

The relation of a material brain and an immaterial consciousness poses a problem. They are very different things, and it is not clear how they would be related. Philosopher David Chalmers called this the "hard problem" of epistemology and consciousness research. He described it as follows:

> When we see, for example, we *experience* visual sensations: the felt quality of redness, the experience of dark and light, the quality of depth in a visual field. Other experiences go along with perception in different modalities: the sound of a clarinet, the smell of mothballs. Then there are bodily sensations, from pains to orgasms; mental images that are conjured up internally; the felt quality of emotion, and the experience of a stream of conscious

thought. What unites all of these states is that there is something it is like to be in them. All of them are states of experience. [34]

"To be in such states" is to experience the qualitative flow of sensations that make up human consciousness. Being in these states is fundamentally different from observing states of the brain. How could a material brain produce an immaterial consciousness?

The hard problem contrasts with the "easy problems" of consciousness research. For example, our ability to discriminate, categorize, and react to environmental stimuli is a relatively easy problem because in principle it could be solved in reference to neural and artificial computational mechanisms. When our brain engages in visual and auditory information processing, we have visual and auditory experiences. The same holds for our understanding of the way the nervous system accesses its own internal states. But the hard problem remains. How can network of neurons that process nerve signals conveyed through the senses produce felt qualitative experience? Philosopher Jerry Fodor said that "nobody has the slightest idea of how anything material could be conscious, nor does anybody even know what it would be like to have the slightest idea about how anything could be conscious."[35]

The hard problem is indeed hard. But it is a problem only if posed under the materialist or the dualist assumptions. (It is not a problem for idealists, for in a consistent idealist perspective the brain is not material: it is itself an element of consciousness.) The relationship of a material brain and an immaterial consciousness is not a real but merely a pseudo problem. Brain and mind are not related through any form of cause and effect: they are categorically different. They are not categorically different *things* but categorically different *aspects* of things. Of the same things—of all things.

34 Chalmers, David J. 1995. "The puzzle of conscious experience" *Scientific American* 273 (December).

35 Fodor, Jerry A. "The big idea." 1992. *New York Times Literary Supplement*, (3 July).

The dual aspect theory

The dual aspect theory overcomes the classical mind–body problem by providing an alternative to the doctrines of materialism, idealism, and dualism. Its principal merit lies, however, not in overcoming a recalcitrant problem but in paving the way to a more adequate understanding of the place of mind in the cosmos.

Mind, in terms of the dual–aspect theory, is not an element *in* the cosmos; it is an *aspect* of the cosmos. The world is not entirely material, nor is it purely mental. Matter and mind are equally descriptive of the reality of the world: they are different but complementary aspects of it.

Not only human beings, and not even just living organisms, *all* things that arise and evolve in the cosmos have two complementary aspects: a material (more exactly, matter–like) aspect, and a mental (or mind–like) aspect.* The matter–like aspect is public; it is available to all observers beyond the experiencing subject. The mind–like aspect is private, available only to the subject. The two aspects are complementary in the sense suggested by Niels Bohr in regard to quantum phenomena.

According to Bohr quanta have both a wave– and a corpuscular–aspect. These aspects are complementary insofar that either the one or the other is manifest but not both. Which is manifest depends on the kind of questions we ask and the kind of experiments we perform.

The same principle applies to all things in the world. Their matter–like and their mind–like aspect become manifest under

* We shall use the expression "matter–like" instead of "material" because matter as such does not exist in the universe. In the final count material things are only *matter–like* things: integrated energies that reflect light and resist touch, but are not hard and solid substances, as classically conceived. Similar caution is indicated in regard to the use of the term "mental." The indicated aspect of things is not what is classically meant by "mental"—it is an aspect that recalls one's intimate experience of consciousness. It is an aspect that is thought–like and feeling–like, ineffable and charged with meaning. But in the final count this quality of existence must be experienced to be known—it defies description in ordinary language.

different conditions. The matter–like aspect appears for the external observer; and the mind–like aspect is displayed for the introspecting observer. Both aspects are fundamental features of things in the world.

Viewed from the outside, human beings and all things in the world disclose their matter–like aspect: atoms, molecules, cells, organs, and organ–systems embedded and entangled in a multicellular organism. In that perspective humans have a brain and nervous system, but they do not exhibit a consciousness. On the other hand viewed from the inside—in the introspective perspective—the experiencing subject exhibits a consciousness: in fact, exhibits nothing but a consciousness. Its own body, and the whole world, are elements in this consciousness.

For the dual aspect theory the hard problem of relating matter and mind does not exist. Matter and mind are not different things related in a causal way one to the other; one does not cause or produce the other. They are different aspects of things—of the same things and of all things. They are not *related*: they are *one*.

The dual–aspect theory lifts the mind–body problem to the level of cosmology. The things that exhibit both mind–like and matter–like aspects are manifestations of the fundamental Akashic dimension of the cosmos. If so, and if the things we observe have matter–like as well as mind–like aspects, then the A–dimension of the cosmos must have the power to manifest matter–like as well as mind–like entities. It must be matter–like as well as mind–like in itself.

Neither *physis* nor *psyche* are the unique reality of the world. The world is *psychophysical*.

Accessing the psyche of the world

Can we experience the mind–like aspect of the world? Doing so would be like entering into communion with God or the spirit or soul that pervades the world. It would also correspond to the frequently voiced claim (or aspiration) "reading the Akashic Record."

Here we enter the domain of religion and spirituality and need to tread cautiously. There is ground under our feet, however: the Akasha paradigm offers a meaningful point of departure.

We have noted that viewing things as external observers brings to the fore their matter–like aspect. In this perspective humans, as other living organisms, are physico–chemical and biological systems. They are formed by physical and physiological interaction with the M–dimension, and are "in–formed" by the subtle signals that reach them from the A–dimension.

The introspective viewpoint, however, exhibits the mind–like aspect things. But this perspective is available only to the experiencing subject. On introspection we experience the world as part of the flow of consciousness that accompanies us from cradle to grave—and without a shadow of a doubt we cannot experience anything else.

Do we experience the *psyche* of the cosmos when we experience our consciousness? Spiritual people would say yes. Their affirmation is warranted by the consideration that the Akasha is a holographic medium. Everything in the Akasha, everything that *is* the Akasha, is present at all its points. The Akasha is therefore present in our consciousness; present in its entirety. When we introspect on our consciousness we encounter more than a cranium–enclosed local mind. We encounter the nonlocal consciousness of the world.

Metaphorically we can say that our individual consciousness "reflects" the mind–like aspect of the world. This is an old insight, beautifully illustrated inter alia in the Hindu legend of Indra's Net. The net of the Vedic God Indra hangs over his palace on Mount Mera which is, according to Vedic cosmology, the axis of the world. The net is made up of a string of jewels, and each jewel reflects all the other jewels. Thus it reflects the whole net. The same applies to our consciousness. It reflects the consciousness of all things in the cosmos, for it reflects the mind–like aspect of the Akasha.

Prophets, shamans, and spiritual people have always told us that we can achieve oneness with a higher mind or spirit by entering into ourselves. The way to do so, they advised, is by deep prayer and meditation.

They are right. When we descend into the deep regions of our own consciousness we enter into an altered state that brings us into nonlocal communication with the mind–like aspect of the otherwise hidden but ultimately real A–dimension of the cosmos.

The oneness of individual consciousness with the mind–like aspect of the Akasha has far–reaching implications. As Akashic cosmology maintains, the manifest world evolves and devolves in the successive cycles of the metaverse. In each cycle the individual entities of the M–dimension come into being and disappear, but the Akasha does not undergo cycles of emergence and dissolution: it is conserved across the cycles and gains in articulation and power to in–form the manifest world. The matter–like aspect of systems emerges and fades—fluctuates between higher and lower forms of organization—whereas the mind–like aspect of the systems is conserved. It is conserved as it is in union with the mind–like aspect of the Akasha. Consequently our individual consciousness does not vanish. It continues to be present in the holographic domain of the cosmos even when the system with which it was associated vanishes. *Psyche* is associated with *physis*, but is not generated by *physis*. It is an enduring attribute of the psychophysical cosmos.

The age–old and quasi–universal cultural belief in the survival of a spirit, soul, or consciousness is not mistaken. And the considerable body of evidence for the continued presence of mind and consciousness in the world—after–death communication (ADC), instrumental transcommunication (ITC), electronic voice phenomena (EVP), near–death experiences (NDEs), and credible reports of rebirth and reincarnation, among others, are not without substance. The A–dimension conserves the mind–like aspect of the things that arise in the M–dimension. The Akasha is not only a coherence–generating interconnecting medium; it is also a cosmic mind with universal memory.

Annex

THE AKASHIC (R)EVOLUTION IN PHYSICS

Four Hypotheses

According to the Akasha paradigm the phenomena we observe are manifestations of an in–itself unobservable but real dimension of the universe, the cosmic matrix we call the Akasha. The hypotheses presented here offer support for this claim: they demonstrate that the basic parameters of the physical universe can be mathematically deduced from the Akasha, identified here as the Unified Quantum Field, the Ether, or the Aether.

Annex I.

The Universal Quantum Field

Foundations for the Unified Description of Basic Physical Quantities

Peter Jakubowski

This study gives a quantitative demonstration that the whole of contemporary physics, with all its equations, units, and constants, can be redefined and unified on the basis of the Universal Quantum Field, an interpretation of the Akasha, defined by only two "classical" factors: the Planck–constant (h), and the elementary electric charge (e).

The UQF is observable only through its quantized fluctuations, named fluctuons, which move across the field, exist for some time, and then disappear. The manifest world is a composition of fluctuons.

The basic equations of contemporary physics emerge as relations among the "unified family" of the physical quantities generated by deduction from the Universal Quantum Field. The UQF does not need definition by observed and experimentally measured values beyond the Planck constant and the elementary electric charge. The unified family of all physical quantities "produces" all physical equations as simple relations between these quantities. The equations appear in their quantized, relativistic and matter–independent form.

Introduction

Our knowledge of the physical properties of the hidden dimension of the universe (as described in Chapter 5) is just beginning to assume a solid place in contemporary science. The physical properties of "the Akasha" can serve as the basis for describing all phenomena manifest in the universe.

In order to develop the physics of the Akasha, we demonstrate how a purely physical Universal Quantum Field (UQF) can re–define and unify the whole of contemporary physics, with all its physical equations, units and constants, supported only by two "classical" values: the Planck constant h, and the elementary electric charge *e*.

We begin by describing the UQF. This field is observable only through its quantized fluctuations (named fluctuons), which move across the field, exist for some time, and then disappear again. The observed world is but a composition of fluctuons.

This is correct, but it is not enough. We need a complete physics that can give us a description of the various physical quantities of the known world.

The physics of the UQF

How can we characterize a fluctuon? Let us consider it as an object having its quantum wave vector **k** and quantum vector of propagation–velocity **c**. Let us call these the fundamental physical quantities, **k** and **c** (we note all vector physical quantities in bold letters). What is the relation of these quantities to each other? What other physical quantities can we define on the basis of **k** and **c**? And what other possible relations exist between the additionally defined physical quantities? Are all of them known in mainstream physics? In order to answer these questions let us introduce the following two–dimensional diagram which is a chess–board, where there is a free place for every basic physical quantity we wish to define and correlate with each other.

Figure 1 shows the central part of the Unified Family with the quantum wave vector **k** and the quantum velocity **c** placed directly

R\C	-1	0	1
-1			**c** velocity
0	**k** wave vector	°1 universal unity	
1			

Fig.1. Fundamental physical quantities, the vectors **k** and **c**, in relation to the scalar universal unity, in the central position of the Unified Family of all physical quantities.

near the "universal unity" of the family. This unity–quantity plays an important role in the whole family, because many of the traditionally used physical quantities are in simple reciprocal relationship with each other thus we need not define them separately. For example, we are able to define immediately the quantum length **r** as a vector reciprocal to the wave vector **k**, $\mathbf{r} = (1/k)\hat{\mathbf{u}}$, where $\hat{\mathbf{u}}$ is a corresponding unit vector, and put it in the position opposite to **k**, on the right side of the universal unity, as shown in figure 2.

Our first question is: What is the relation of the fundamental physical quantities **k** and **c** to each other? The only such relation can be written as an outer product of these two vectors. It defines the quantum frequency of the fluctuon as a bivector »f«[36]:

$$\text{»f«} = \mathbf{k} \wedge \mathbf{c}. \qquad (1)$$

Equation (1) is not only the first physical equation of the unified description of the observable world; it is in principle the single

36 The bivector is an oriented plane, here a plane over **k** and **c** with a determined circulation along **k** and **c**. It is noted in Figure 2 and in all following figures with a double line above the bivector quantity.

R\C	-1	0	1	
-1		$\overline{\overline{f}}$ frequency	**c** velocity	
0	**k** wave vector	$^{o}1$ universal unity	**r** length	
1		$\overline{\overline{t}}$ period		

Fig.2. Central part of the Unified Family of all physical quantities. The bivector of the quantum frequency »f« is defined with Eq.1 (see above), the quantum period »t« is its reciprocal, »t«= (1/f)»u«, where »u« is a unit bivector, and the vector length **r** is reciprocal to the wave vector **k**, **r** = (1/k)**û**, where **û** is a unit vector.

basic equation of the UQF, the Universal Quantum Field. It describes every single fluctuon of this field. Theoretically, nothing more is necessary for the description, because our observed world consists exclusively of fluctuons.

However, we want to know something more about the world. And if so, we need to answer also the second question: What other physical quantities can we define on the basis of **k** and **c**?

Two examples of such additional quantities are already present in Figure 2: they are the reciprocals of the quantum wave vector **k** and the quantum frequency »f«. We have named the reciprocal of the wave vector **k** quantum length **r**. This is also the quantum size of the fluctuon. The reciprocal quantum frequency »f« has been called quantum period in Figure 2 (and defined as »t«=(1/f)»u«, where »u« is a unit bivector), but it means also a quantum time, which is bivector, and not a scalar, as in conventional physics. It is our first drastic difference to the classic understanding of the observed world. In the unified quantized world we have not the classical, linear flow of time, from the past to the future. The universal quantum time has

R\C	-1	0	1	2	3	4
-1		$\bar{\bar{f}}$ frequency	**c** velocity			
0	**k** wave vector	$\overset{\circ}{1}$ universal unity	**r** length	$\bar{\bar{A}}$ area	**p** momentum	$\overset{\circ}{J}$ action
1		$\bar{\bar{t}}$ period				

Fig.3. The consecutive multiplication of the universal unity with quantum length **r** results in bivector quantum area »A«, vector quantum volume **V** (equivalent to quantum momentum **p**) and finally in the scalar quantum of action J.

always the sense of a characteristic period of some corresponding quantum system, even if that system is a milliard light–years big. Quantum time does not flow, it circulates.

We now consider what further physical quantities we can define on the basis of **k** and **c**? We enlarge the plane of Figure 2 to the right and thus obtain Figure 3.

As shown here, by repeating twice the multiplication of the universal unity with the quantum length **r**, we obtain the bivector quantum area »A« (»A« = $\mathbf{r}_1 \wedge \mathbf{r}_2$). Multiplying it with **r** again gives the quantum volume **V** of the fluctuon (**V** = »A«***r**, where * means eigen–product), which is equivalent to the quantum momentum **p**. Finally we multiply it with **r** and obtain the scalar quantum of action J, J = **p*****r** (= »A«*»A«).

But how do we know that the fluctuon volume **V** is equivalent to its momentum **p**? We need to put this to the test. If it works with the other members of the Unified Family, it is correct. And if we have chosen the right place for the momentum, then we have also have the right place for action. This is because the classical and

R\C	-1	0	1	2	3	4
-1		$\overline{\overline{f}}$ frequency	**c** velocity	$^{\circ}\Phi_f$ frequency flux	**F** force	$\overline{\overline{W}}$ energy
0	**k** wave vector	$^{\circ}1$ universal unity	**r** length	$\overline{\overline{A}}$ area	**p** momentum	$^{\circ}J$ action
1		$\overline{\overline{t}}$ period		$^{\circ}m$ mass		

Fig.4. The position we assumed for quantum mass m is shown to be in proper relation to the already defined physical quantities, **p**, J, **F** and »W«.

the quantum equivalence of action in regard to angular momentum is given above: J = **p*r**.

Let us test the position for linear and angular momentum. Consider what physical quantities could fill row –1 in Figure 3. In light of Newton's classical physics, we can suppose that multiplying momentum **p** with frequency »f« would give us the correct position for the quantum vector of force **F**, **F** = »f«***p**. Therefore we place force **F** on the intersection of column 3 with row –1.

We also know that force times length gives energy. If we place energy W[37] to the right of force (in column 4 of row –1), then we have to define it as the bivector »W« = **F** $\wedge$ **r**. Finally let us note that the physical quantity in column 2 in the same row must be the scalar quantum flux of frequency Φ_f, because it equals the (two–dimensional) quantum frequency »f« times the quantum area »A«, Φ_f = »A«*»f«.

In order to complete the dynamic plane of the Unified Family, we now need the correct position for quantum mass m. This is indicated in Figure 4.

A brief analysis of the known relations between quantities that are already defined proves the correctness of the proposed position for scalar quantum mass m.

37 We have to use the symbol W for work to note energy, because in the Unified Family the standard symbol E must be left to denote electric field strength.

One step further along the diagonal direction between the rows and columns in Figure 4 leads us from the universal unity to quantum velocity **c**. This suggests that every step along this or a parallel direction means multiplication with the quantum velocity **c**. This gives the following relations: m***c** = **p**; **p** $\wedge$ **c** = »W«, and m*($\mathbf{c}_1 \wedge \mathbf{c}_2$) = »W«.

We now ask, what do we know about the units of the here defined physical quantities? If the quantum mass is correctly included in the Unified Family, then we have the following definition of its unit, the kilogram: kg = m^2s (because of the relation in the row 1: »t«*»A« = »A«*»t« = m).

Let us test this definition. In classical physics the unit of force should be kg*m/s^2. Using our definition of kilogram, we can deduce that N = m^2sm/s^2 = m^3/s. Due to the relation **F***»t« = **p** (in column 3), we obtain as the unit of momentum the quantity m^3, and this is also the unit of volume **V**. It follows that the quantum volume **V** and quantum momentum **p** are equivalent physical quantities, and as such they occupy the same place in the Unified Family.

Now, the unit of energy is Joule: J = Nm = kg*m^2/s^2 = m^2s*m^2/s^2 = m^4/s. Our column 4 gives the relation for quantum action J = »W«*»t«. It gives for the unit of action the combination m^4, the unit of volume multiplied by meter. Also the position for quantum action has been chosen correctly. This is important. It means that we can take the well known value of the Planck constant (J_u = h = 6.626076×10^{-34} Js) and calculate with its fourth root the universal value of quantum size in the observable world: $r_u = 5.073575\times10^{-9}$m; ($r_u^4$ = h).

We have confirmed the correctness of the positions of all physical quantities in Figure 4. Let us thus take the most important discoveries up to this point. First, the quantum frequency is a bivector; it describes circulation, rotation, or spinning of a fluctuon. Second, the quantum time does not flow, it circulates; it has always a sense of some shorter or longer period. Third, the quantum of energy is not a classical scalar, it is a bivector. It means, that a quantum of energy contains all the information that in traditional physics we had to add by means of the properties of the energy carrier, the photon. Every quantum of energy carries its own spin and spatial orientation. In Unified Physics we do not need both energy quanta and photons; we do not need photons at all, as the unified quanta of energy are their own energy carriers. Moreover the universal "building blocks" of the observed world are not atomic or molecular objects; they are nano–size fluctuons.

Finally, we have discovered something distinctly promising. We no longer need the kilogram standard anymore. We are able to calculate the mass of a quantum object from its characteristic, quantum period and its size (or from any two other of its physical quantities).

Let us now complete the Unified Family. First we complete its dynamic plane, as shown in Figure 5.

Figure 5 shows all the physical quantities that constitute the dynamic basis of the Unified Family of physical quantities. As we already know, some of these quantities are simply reciprocals of the other, as shown already in Figure 4. Quantum Laplacian »Δ« is simply the reciprocal of quantum area »A« and quantum density **n**

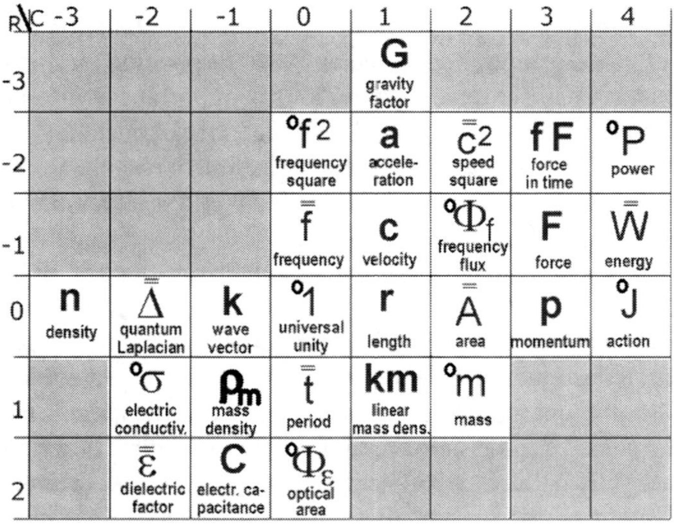

Fig. 5. The most frequently used physical quantities belonging to the dynamic plane of the Unified Family

is reciprocal of the quantum volume **V** (alias momentum **p**). The quantum mass density ρ_m is the reciprocal of the quantum velocity **c** (because of the relations $\rho_m = (m/V)\hat{\mathbf{u}}$ or $\rho_m * \mathbf{V} = m$); another discovery resulting from our definition of unified mass.

The position of the physical quantities that we call frequency square and speed square may be obvious as well. They are defined correspondingly as f^2 = »f_1«*»f_2« and »c^2« = $\mathbf{c}_1 \wedge \mathbf{c}_2$. They are often used in traditional mathematical physics. At the moment it is enough to note that the frequency square has as its reciprocal the optical area (used in some specific optical equations), and the speed square is the reciprocal of the so–called dielectric factor »ε« (because of the well–known relation »ε«*»c^2« = 1). This relation means, however, that the dielectric factor, like its flux (the optical area) belongs to the dynamic plane of the Unified Family. Also the electric capacitance resulting from a multiplication of the dielectric factor with the quantum length **r**, **C** = »ε«***r**, belongs to this plane. Though the names "dielectric factor" and "electric capacitance" suggest something new, both these quantities have been traditionally defined purely dynamic, and not as electrodynamic, quantities.

In order to properly introduce the electric conductivity in the Unified Family, we need to define its electrodynamic plane. We will construct it below. But first we need to deal with some important physical quantities in the upper part of the dynamic plane in Figure 5.

It is relatively easy to find the right position for quantum acceleration **a**. It belongs to the position defined by a multiplication of the quantum velocity **c** with the quantum frequency »f« , or by the multiplication of the square of frequency with the quantum length **r**, **a** = »f«***c** = f^2***r**. It is also clear that quantum power should lie above energy, because of its definition as the amount of energy "applied" or "used" during a specific period of time, P = »f«*»W«.

The other defining relations are P = **F*c** = »f«***F*r**. The latter defines also the position of the physical quantity **fF** (I have called it force in time, because it describes the change of the quantum force during the corresponding quantum period »t« (as the reciprocal of »f«). The units of acceleration (m^2/s) and power (W = J/s) arise directly from the above definitions.

The last physical quantity at the top row of Figure 5 is the next important finding of Unified Physics. This quantum–vector quantity is called the gravity factor **G,** because it plays the same role in the definition of the unified force as the gravity constant does in the Newtonian force of gravity. However, the unified gravity factor is not just a constant; it is an "ordinary" physical quantity that describes the temporal change of quantum acceleration, **G** = »f«***a**. Its unit is m/s^3. The gravity factor becomes zero for all motion with constant acceleration. This discovery radically changes our understanding of the relation of gravity to antigravity.

Next, before we calculate the universal value of the gravity factor and the universal values of the other physical quantities, we have to construct the electrodynamic plane within which we locate them.

Because of known historical reasons, the development of the classical physics of magnetism and electricity occurred independently of each other and also independently of the other parts of physics, such as dynamics and kinematics. It would be thus quite understandable that the physical quantities introduced into each of these fields of physics can be defined independently of the other physical quantities. Fortunately for us, however, our scientific predecessors were great scientists who were intuitively aware of the unity of nature. Besides, they were also smart enough to understand not only their own field of physics, but the other fields as well. Thus the system of electrodynamic physical quantities and units, developed during the 18th and 19th centuries, is compact and self–consistent. We can use the multiplication rules we use in the dynamic plane to calculate all possible electrodynamic quantities, i.e., use the UQF's two fundamental quantities, the quantum wave vector **k** and the quantum velocity **c**. The single problem that remains to be solved, using intuition, experience and some luck, is to find out the proper numerical relationship between both planes of the Unified Family, the dynamic and the electrodynamic. This is mapped in Figure 6.

When we compare the Figure 6 with Figure 5, we find that the electrodynamic place of the universal unity has been occupied by another scalar physical quantity, namely magnetic induction B, which is equivalent to the planar density of the electric current j.

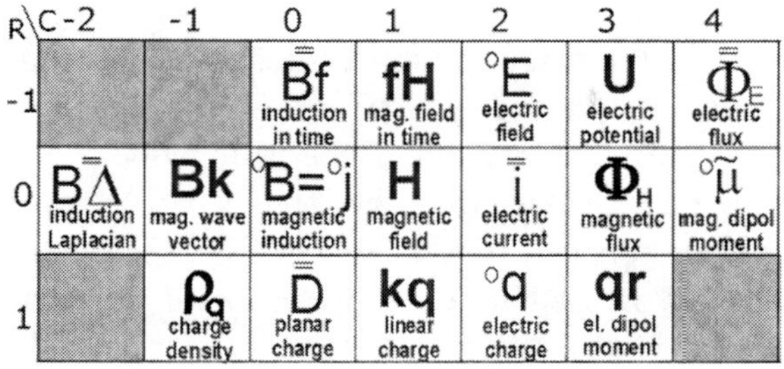

Fig. 6. All useful physical quantities belonging to the electrodynamic plane of the Unified Family *(The numbering of the columns and rows is the same as in Figure 5.)*

Like the universal unity, also this scalar quantum physical quantity, B = j, is a universal constant, independent of the material state (we define the matter–dependence of physical quantities in Figure 7). This is just a numerical factor that transforms the corresponding part of the dynamic plane into the electrodynamic plane. For example, comparing the actual positions of the magnetic field **H** (in Figure 6) and the quantum length **r** (in Figure 5) we can write directly **H** = B***r** (or **H** = j***r**). Similarly, the quantum electric current »i« equals: »i« = j*»A« and the planar charge (or electric induction) »D« is equal to »D« = B*»t« (better known as Maxwell's displacement current: j = »f«*»D«). As a result of our definition of the quantum mass, we have a new relation between quantum mass and electric charge: q = j*m, which gives us another way to find the numerical factor that connects the two planes. This factor is the quotient of the quantum electric charge in regard to the quantum mass of an arbitrary fluctuon, j = q/m. If we estimate the electric charge of a fluctuon, we immediately obtain also its mass. And the estimation of the quantum mass determines also the electric charge.

In order to complete the unification of all physical quantities we need to find the expression of the electric charge in reference to the fundamental quantities **k** and **c** (or through their derivatives »f« and **r**). I found that the square of the electric charge is a member of the dynamic plane. It can be located on the intersection of row 1 and column 3, next to the quantum mass: $\mathbf{q}^2$ = m*$\mathbf{r}$ = »t«*$\mathbf{r}^3$. This location is the origin of some interesting relations, for example, $\mathbf{q}^2$*»f« = $\mathbf{p}$, $\mathbf{q}^2$*$\mathbf{c}$ = J, or $\mathbf{q}^2$*f^2 = $\mathbf{F}$. The last named is the indicator for the correct place to find the still "hidden" quantity $\mathbf{q}^2$.

The traditional expression for the force generated by two wires carrying electric current is well known: $F = i^2$. In our quantum language it is expressed almost the same way: F = (q*»f«)^2. The only difference is that here the electric current is the bivector »i« and the classic current is defined as a vector quantity. As a consequence, the quantity for the square of the electric–charge is a vector, whereas its classic expression treats it as a scalar. This relatively minute difference has obstructed the complete unification of physical quantities for all these years.

The electric charge is the second physical quantity (after the Planck constant) that has been often and precisely measured: it is $q_u = e = 1.602177\times10^{-19}$C. We thus take this quantity as the second universal value in the Unified Family. With a new definition of electric charge q = (»t«*r^3)$^{1/2}$, we are now able to calculate the universal value of the second fundamental physical quantity. It is $t_u = e^2/r_u^{\ 3} = 1.965526\times10^{-13}$s, and $c_u = r_u/t_u = 2.581281\times10^4$ m/s. This value of the fundamental quantity **c** is a well–known quantity in physics when we realize that the quantum velocity of a given material is equivalent to its quantum electric resistance. The same value as c_u above but expressed in Ohms, has been measured by Klaus von Klitzing in his investigation of the Hall–effect. It has earned him the Nobel Prize in physics in 1985.

The third possible plane of thermodynamic physical quantities should contain only one relevant item, namely temperature. Thus we treat temperature separately and simply add it to the Unified Family of physical quantities. The results are given in the two tables of Figure 7.

The defining relation is given in the left upper corner for each quantity. This definition includes the direct dependence on the material factor μ. Having extracted the material dependence (with different powers of μ) the remaining part of the definition can be expressed through the fundamental universal values (noted with the subscript "u"). The definition of the corresponding unit is given in the right upper corner, and the universal value below

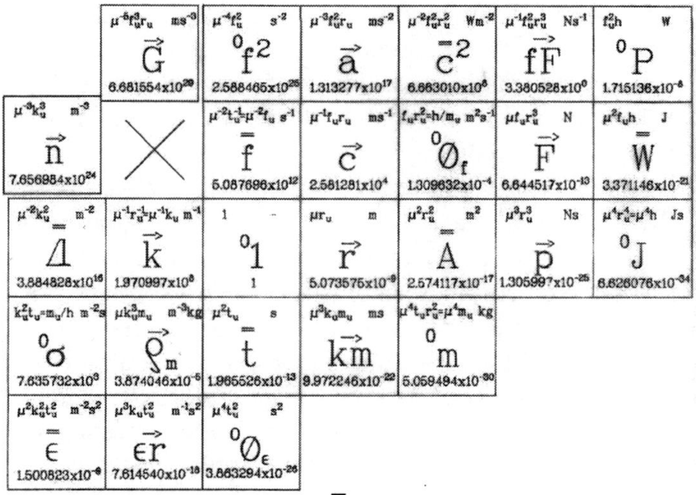

7a

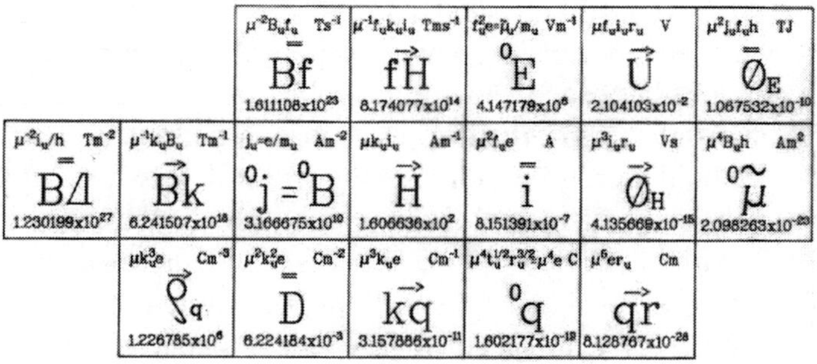

7b

Fig. 7. The complete definition of the Unified Family of all physical quantities.

the symbol of the given physical quantity. The dynamic plane is shown in Figure 7a (note the shifted positions of G and n in relation to Fig.5 – for the picture–size reduction) and the electrodynamic plane is shown in Figure 7b.

One new property of the Unified Family has been demonstrated in Figures 7a and 7b in the left corner of each square. The unified definition of every physical quantity contains a corresponding material factor μ, which takes on the value 1 for the average state of the universal quantum field UQF, a value between 1 and 0 for all possible states of inanimate matter and a value greater than 1 for all possible states of animate matter. (See the References below for a full definition of the quantum spectrum of matter and a detailed discussion of its properties.)

Let us return now to the Unified Family of physical quantities. We are speaking about all physical quantities, but Figure 7 contains “only” 44 such quantities, or 45, including temperature. Where are the other hundreds known to science? The fact is that they are all equivalent to one or another of the chosen “basic” quantities shown in Figure 7. For example, to the quantities equivalent to frequency flux, Φ_f, belong in addition to the quantum of circulation and the quantum of rotation also the diffusion coefficient, elastic force coefficient, and electric resistivity. In addition to the already noted electric resistance, also the energy density and pressure belong to quantum velocity. The Unified Family does indeed define all physical quantities.

As a consequence of its completeness it also produces all possible relations between the quantities. We know them as physical equations, and some of them appear more important than others. Those we hold more important are known as the laws of physics, or even the laws of nature. This is misleading, because nature “works” without any physical equations. The only process I am ready to call a law of nature is the tendency in nature to transfer energy into the part of space where there is an energy deficit; or conversely, to transfer energy out of a region that has an energy excess in relation to its surroundings.

In science, and especially in physics, we have produced a great number of descriptions of phenomena and processes that seem unique and in need of a special explanation. The Unified Family offers explanations for 461 relations. We can read all of them directly from Figure 7. The most famous among them are given in the table following, in reference to the name of their discoverers.

M is here the material factor μ^{-3} (compare corresponding definitions in Figure 7) compensating for historically “created” difference between material dependences of always two electrodynamic quantities on the right side of the equation and a dynamic quantity on its left side.

Discoverer	Concerns	Law
Ampere	electric current	$»i« = \mathbf{r}*\mathbf{H}$
Ampere	current density	$j = \mathbf{k}*\mathbf{H}$
Biot–Savart	magnetic field	$\mathbf{H} = »i«*\mathbf{k}$
Coulomb	electric force	$\vert\mathbf{F}\vert = M*»i«*»i«$
Coulomb	magnetic force	$\vert\mathbf{F}\vert = M*\mathbf{H}*\Phi_{\mathbf{H}}$
D'Alembert	Second time derivative	$f^2 = »c^2«*»\Delta«$
De Broglie	momentum	$\mathbf{p} = J*\mathbf{k}$
Einstein	energy	$»W« = m*»c^2«$
Faraday	magnetic induction	$»Bf« = E*»\Delta«$
Faraday	magnetic field	$\mathbf{fH} = E*\mathbf{k}$
Faraday-Henry	electromotive force	$\mathbf{U} = »f«*\ \Phi_{\mathbf{H}}$
Gauss	electric flux	$q = »\varepsilon«*»\Phi_E«$
Gauss	electric– induction flux	$q = »A«*»D«$
Hooke	elastic force	$\mathbf{F} = \Phi_f*\mathbf{r}$
Lorentz	electric force	$\vert\mathbf{F}\vert = M*q*E;\ (E = q*f^2)$
Maxwell	displacement current	$»i« = \Phi_f*»D« = »f«*»A«*»D«$
Maxwell	charge density	$j = »f«*»D«$
Newton	acceleration	$\mathbf{F} = m*\mathbf{a}$
Newton	dynamic force	$\mathbf{F} = »f«*\mathbf{p}$
Ohm	electric conductivity	$j = \sigma*\Phi_E$
Ohm	electric resistance	$\mathbf{U} = »i«*\mathbf{c}$
Planck	quantum energy	$»W« = J*»f«$
Poisson	charge density	$\mathbf{U} = \rho_{\mathbf{q}}*P$
Poisson	linear charge density	$\mathbf{U} = \mathbf{kq}*»c^2«$
Poynting	radiation intensity	$\vert»c^2«\vert = M*E*\vert\mathbf{H}\vert$
Schrödinger	quantum energy	$»W« = J*»f« = (J^2/m)*»\Delta«$

We see, for example, that the Ohm equation for electric conductivity relates the electric field E from the electrodynamic plane with the electric conductivity from the dynamic plane. By comparing the place of the electric field with its corresponding physical quantity on the dynamic plane we find the scalar quantum flux of frequency Φ_f. On the other side, this frequency flux is reciprocal of the electric conductivity, $\Phi_f = 1/\sigma$. This gives the relation $E = j*\Phi_f$, being the origin of the Ohm's relation. This means that we have properly chosen the "basic" place also for the quantum electric conductivity on the dynamic plane.

Some other physical equations that are not identified with the name of their discoverer are shown in the following table.

Description of the relationship	**Equation**
Acceleration to rotation	$\mathbf{a} = f^2*\mathbf{r}$
Electric capacitance to dielectric factor	$\mathbf{C} = »\varepsilon«*\mathbf{r}$
Electric capacitance to optical area	$\mathbf{C} = »\Phi_{\varepsilon}«*\mathbf{k}$
Electric charge to electric field	$q = E*»\Phi_{\varepsilon}«$
Electric charge to electric current	$q = »i«*»t«$
Electric charge to electric dipole moment	$q = \mathbf{qr}*\mathbf{k}$
Electric field to electric charge	$E = q*f^2$
Energy to magnetic dipole moment	$\lvert»W«\rvert = M*\tilde{\mu}*\lvert\mathbf{H}\rvert$
Energy to power	$»W« = P*»t«$
Energy to momentum	$»W« = \mathbf{p} \wedge \mathbf{c}$
Force to energy	$\mathbf{F} = »W«*\mathbf{k}$
Magnetic dipole moment to circulation	$\tilde{\mu} = q*\Phi_f$
Magnetic field to magnetic flux	$\mathbf{H} = \Phi_{\mathbf{H}}*»\Delta«$
Magnetic field to its own change in time	$\mathbf{H} = \mathbf{fH}*»t«$

Description of the relationship	Equation
Momentum to force	$\mathbf{p} = \mathbf{F}$*»t«
Electric induction to electric charge density	»D« = ρ_q*$\mathbf{r}$
Power to action	$P = J^*f^2$
Power to force	$P = \mathbf{F}^*\mathbf{c}$
Circulation to mass	$\Phi_f = m^*f^2$

In summary, we have shown that the Universal Quantum Field, defined as above, does not need further reference to observed and experimentally measured values than the two (arbitrarily) chosen universal values, the Planck constant h and the elementary electric charge e. The Unified Family of all physical quantities, constructed in order to correctly define all basic quantities hitherto used in traditional physics, "produces" all physical equations as simple relations between these quantities. And it is noteworthy that these equations always appear in their quantized, relativistic and matter–independent form.

In addition, UQF–based unification produces a number of quantitative relations unknown in mainstream physics, that, for example between quantum mass and quantum electrical charge.

References

Jakubowski, P., 2010, *Naturics; The unified description of Nature*, 2nd edition, Books on Demand GmbH, Norderstedt, ISBN: 978–3–8334–6932–9

Jakubowski, P., 2003, *The cosmic carousel of life; Our evolution and our perspectives*, Books on Demand GmbH, Norderstedt, ISBN–10: 3–8330–0402–9

Jakubowski, P., 1992, Alternative Foundation of Physics, *Physics Essays*, **5**, 1, pp. 26–38

Jakubowski, P., 1990, Luminiferous Ether Revived, *Physics Essays*, **3**, 3, pp. 281–283

Jakubowski, P., 1990, Equivalence of Electrodynamics with Dynamics, *Physics Essays*, **3**, 2, pp. 156–160

Jakubowski, P., 1978, AFMR in $KMnF_3$ near T_N and T_C; New experimental results, *Acta Physica Polonica*, **A54**, 4, pp. 397–409

Jakubowski, P., 1976, Best optimized One–Electron Wave–Functions; I. The General Procedure of Optimization; II. Isoelectronic Series of Li, Be, B, and C; III. Direct Examination of Optimization Effectiveness; IV. Ionization Energies of Atoms, *International Journal of Quantum Chemistry*, **X**, pp. 719–746

Website of naturics: www.naturics.eu

Annex II.

The Superfluid Ether as the Basis for Unifying the Fields and Forces of Nature

László Gazdag

This hypothesis describes the Akasha as a superfluid and superdense cosmic medium, a reconceptualization of the late 19th century space–filling ether. In this medium Dirac's negative energy state becomes the superfluid state, and the quantity of negative energy used to leave the superfluid state is the quantity of positive energy exerted in the excitation of the field.

The equations of general relativity describe the flow of this superfluid gravitational field. The field is non–Euclidean; it acts Euclidean only if the given inertial system is at rest or moves with uniform rectilinear movement in relation to the field. The reconceptualized ether is the cosmic matrix underlying physical phenomena in the universe.

The enterprise of grand unification

After in 1865 James Clark Maxwell unified the electric and magnetic fields, in 1979 Abdus Salam, Sheldon Glashow and Steven Weinberg were jointly awarded the Nobel Prize for the unification of the electromagnetic and the weak interactions. The *electroweak* theory was born based on the neutral Z° scalar boson and the electrically charged $W^{\pm}$ vector boson having two electric charges.

This is formally very similar to the theory of pi–meson nuclear forces developed earlier by Hideki Yukawa, where two electrically charged pi–mesons and a neutral pi–meson (pion) transmit the interaction in the field of the atomic nucleus ($\pi^{\pm}$, π°).

Today we know that the pi–meson interaction offers only a phenomenological description, describing the surface of the phenomenon without paying attention to the processes that occur below the surface. The nuclear force is actually a weak emanation of a kind of van der Waals force of the gluonic interaction between the quarks within the nucleons and the outside of the space of the nucleus.

The same can be said about weak interactions. The description based on the Z and W–bosons is only a phenomenological description. The real mechanism is the weaker effect of the quasi–magnetic phenomenon that accompanies the gluonic interactions within the nuclear space outside the nucleon. Thus the weak interaction is not an independent interaction but a magnetic phenomenon that accompanies the strong interaction. Since this description of the electroweak interaction is also only phenomenological, it is not possible to set forth the project of unifying the known forces.

A further unification is possible, on the other hand, based on the quantum vacuum viewed as a superfluid universal field. This is a new interpretation of the concept of the ether, which in is original form has been eliminated from the physicists' world–picture at the beginning of the 20th century.

Physics in the 19th century operated with the concept of ether as the medium that carries slight. Thus, though this has not been explicitly recognized, the ether should be a superfine medium, since there is no friction between it and the planets—the motion of the latter is not slowed by movement through it. Light would logically have to be a vibration of the ether. The fact that light is a transversal (elecromagnetic) wave was a problem, since transversal waves propagate only in solid media. The ether should be superfine and should also be a solid. In the end, when the negative results of the Michelson–Morley experiments came to light in 1887, the theory of the ether was discarded.

The ether as a fundamental medium

The ether must be considered not as a material medium producing friction, but as a superfluid, frictionless, and yet superdense universal medium.

We now know that there are mediums that are at the same time superfine and superdense. This is the case in regard to helium, which can be superfluid and can also be a superconductor.

In 1908 Kamerlingh Onnes (1853–1926) was able to cool helium to 4.2 degrees Kelvin at which point it became fluid. In 1911 he produced superfluid helium, that is, helium that flows without friction at 2.17 degrees Kelvin. Also in 1911, he discovered superconductivity, the flow of a current without resistance in conductors cooled to almost the absolute zero of temperature.

Helium atoms in superfluid helium flow without friction, and all bodies immersed in that medium move without friction. Onnes observed that helium conserves superfluidity up to 26 MPa (Mega Pascal).

Pjotr Kapitsa (1894–1984) spent many years studying the superfluidity of helium, and was awarded the Nobel Prize for his work. Kapitsa noted that waves of different velocities can propagate in superfluid helium. (We know that the propagation velocity of a wave in a given medium is always constant. What can change is the wavelength (and its inverse, the frequency of the vibration) and the amplitude, which is the intensity of the vibration producing the wave.) The finding that waves with different velocities can propagate in superfluid helium is explained by the fact that the waves have different energy and motion (impulse) levels, since there are no collisions between the helium atoms that would equalize the energies (i.e. produce energy dissipation) as in classical gases and liquids. The dense supraconductor behaves as a vacuum for electrons: they do not interact with it and thus flow without resistance.

The universal background field known as the quantum vacuum appears to be superfluid yet dense. This is, in fact, the implication of the physics of the *Dirac sea.*

The Dirac sea

Paul Dirac was searching for a linear form of the Hamiltonian of the relativistic energy of the electron. For a potential free field we have the following equation

$$H = \sqrt{m_0^2 \cdot c^4 + c^2 \cdot p^2}\,.$$

When we introduce the equation

$$H^2 = E^2 = m_0^2 \cdot c^4 + c^2 \cdot p^2 =$$
$$= \left(\alpha \cdot m \cdot c^2 + \beta \cdot p \cdot c^2\right)^2$$

and substitute the energy and impulse operators, we get the following equation for the Schrödinger wavefunction Ψ:

$$\alpha_{xzy} \cdot \nabla\Psi + \frac{1}{c} \cdot \partial\frac{\Psi}{\partial t} + i \cdot \beta \cdot \frac{2\pi mc}{h} = 0$$

where α and β are not scalar but matrix, and through elimination can be identified as spin operators.

The Dirac–equation produces quantities greater than $m_0 \times c^2$ and smaller than $m_0 \times c^2$ for the kinetic energy of the free electron. Thus the concept of negative energy and negative mass was created. Dirac thought that all levels of negative energy in the universe are already completely filled without leaving any empty space or void, and that this was the reason why the negative energy domain is not observable. (Interestingly, the superdense "void–free" negative energy field resembles a superdense quantum vacuum field.)

At the same time Dirac thought that this field was made up of the negative energy counterparts of the known particles, for instance, electrons. By inputting energy to them, they would be turned into positive energy (and positive mass) particles.

If a photon having $h \times \gamma = 2 \times m_o \times c^2$ energy collides with a negative energy electron, it produces a $+m_0$ mass and an $e = +m_o \times c^2$ energy "normal" electron with negative charge, while part of the inputted energy is lost, since it is used for lifting the electron from the negative energy state of the vacuum.

When the electron is lifted from the negative energy vacuum field, it leaves behind a hole which is a positive mass electron with a reversed positive charge. This is the positron. The opposite of this process produces annihilation; here an electron emits a photon and simultaneously "sinks" into the negative energy sea filling the negative–energy hole. The positron and the electron are both annihilated.

Let us now interpret the Dirac–equation on the basis of the characteristics of a superfluid and superdense vacuum field.

The concept of "negative energy" means that an energy input is required for the particle to leave the superfluid state and achieve a state where it is capable of interaction. (Superfluidity indicates a state where interaction is not possible.) The Dirac equation suggests that the vacuum is a superliquid boson field that fills space with quasi–singularities and swarming virtual particles without leaving any void.

Dirac found an equation that complied with the requirements of the theory of relativity and gave the spin of the particle. However, the solution of this equation, or more accurately, one of it solutions, resulted in *negative* energy and negative mass. (The result for the kinetic energy of the electron was not only $+ m \times c^2$, but also $- m \times c^2$.)

This problem is resolved when we assume that there are particles with negative energy and negative mass (i.e., with negative energy levels), but that we cannot observe them because the negative energy levels in the vacuum are completely filled.

The ether as a superfluid vacuum field

Note the functional analogy between Dirac's void–free undetectable sea and the superliquid and superdense vacuum continuum proposed above. When we postulate a superlufid and superdense vacuum, the correct expression of the negative energy state becomes "superfluid state," and the quantity of negative energy used to leave the superfluid state becomes the quantity of positive energy exerted in the "excitation" of the vacuum.

Now we can reinterpret the equations of general relativity as referring to the flow of a superfluid graviton field in the vacuum, which becomes a physical expression of space–time.

A new concept of the ether

Einstein arrived at the famous tensor equations of the general theory through the variation of the action integral of the gravitational field and gravitating matter. Thus formally he had arrived at equations of motion, but viewed them as field equations.

$$R_{ik} - \frac{1}{2} \cdot g_{ik} \cdot R = \frac{8 \cdot \pi \cdot G}{c^4} \cdot T_{ik} \quad 1.$$

The question is whether this interpretation is correct in view of its consequences. Is it not the case that the similarity in form between the relativistic field equations and the classic equations of motion suggests a deeper correlation?

$$\delta(S_m + S_g) = \frac{c^3}{16 \cdot G \cdot \pi} \int \left(R_{ik} - \frac{1}{2} \cdot g_{ik} \cdot R - \frac{8 \cdot \pi \cdot G}{c^4} \cdot T_{ik} \right) \cdot g_{ik} \cdot \sqrt{-g} d\Omega = 0 \quad 2.$$

Einstein varied the g_{ik} metric tensor, but according to David Hilbert, g_{ik} corresponds to the gravitational potential tensor. Here the potential is a quantity with the velocity dimension m/s. Hence the gravitational potential is the flow velocity of the graviton field at the given point. Field intensity has the acceleration dimension m/s^2, that is, it can be considered the rate of acceleration of the field. We know that if there is acceleration, the state of superfluidity ends and macro effects appear.

The gravitational force is essentially the mass acceleration vector.

$$\dot{F} = m \cdot \dot{C} \quad 3.$$

Therefore, the Einstein–equations describe actually the flow of the superfluid gravitational field. If there is a drain at a given point of the field (that is, if the right side of the equation is not equal to zero), the flow of the field accelerates and the trajectory curves. The field is not Euclidean. If the field at the given inertial system is at rest or moves with uniform rectilinear movement, the field is Euclidean. Thus the structure of the field is determined by the flow characteristics of the graviton field.

Field intensity is the acceleration of the field caused by absorbing particles of matter; and in consequence of this acceleration we have the termination of superfluidity and the emergence of macro effects and gravitation. The general relativity equations can be reinterpreted accordingly.[36*]

36* *Reinterpretation of the General Relativity equations*

$$T_{ik} = \frac{2}{\sqrt{-g}} \left(\frac{\partial \sqrt{-g} \cdot \Lambda}{\partial g^{ik}} - \frac{\partial}{\partial x^1} \frac{\partial \sqrt{-g} \cdot \Lambda}{\partial \frac{\partial g^{ik}}{\partial x^1}} \right) \quad 4.$$

where $\Lambda = T - U$ is the Langrage density of the material system, since the integral $L = \int (T - U) dV$ is the Lagrange function.

$$R_{ik} - \frac{1}{2} g_{ik} \cdot R =$$

$$\frac{1}{\sqrt{-g}} \left(\frac{\partial \left(\sqrt{-g} \cdot C\right)}{\partial g^{ik}} - \frac{\partial}{\partial x^1} \frac{\partial \left(\sqrt{-g} \cdot C\right)}{\partial \frac{\partial g^{ik}}{\partial x^1}} \right) \quad 5.$$

C can be expressed using the Christoffel–symbols:

$$C = g^{ik} \cdot \left(\Gamma^m_{il} \Gamma^l_{km} - \Gamma^l_{ik} \Gamma^m_{lm} \right) \quad 6.$$

C is an expression related to the curvature of the time–space, since the Christoffel–symbol "appears" in the course of the parallel transport of the A^i vector in curved frame of reference.

$$\partial A^i = -\Gamma^i_{kl} \cdot A^k dx^1 \quad 7.$$

where A^i is the transformation of the A^i vector.

Toward grand unification based on the superfluid vacuum field (physical space–time or ether)

Based on the superfluid quantum vacuum as physical space–time or ether, Grand Unification becomes possible.

The structure of space–time within the nucleus, or—what is equivalent to it—the trajectory of quarks, is determined by gluonic interaction; electromagnetism and gravitation are negligible. Beyond the nucleon field but still within the nucleus, the determining force is the gluonic force, though it is significantly weaker. (This weakened radiation is the classical nuclear force.)

Outside the nucleus gluonic interactions decay rapidly in consequence of the color charge of gluons. At this level the gluons do not encounter quarks, and as a result of their interactions with one another the gluon field becomes chaotic. But it is at the same time neutral from the point of view of the interactions: space–time remains superfluid.

Inside the electromagnetic interactions encompass the space–time field. The structure of space–time is determined by the interaction between electrons and protons, as gravitation is negligible.

Electromagnetic interactions are determinant also at the level of molecules. However, beyond the level of atoms and molecules electric interactions fade, since protons and electrons shield each other. Naturally, in the event of ionisation, the charges become weaker and electromagnetic interactions can re–emerge as determinant of the structure of space–time. This however is a temporary, local state. In the interaction between atoms and molecules

---

Please note, that according to Hilbert the interpretation of the Γ^{i}_{kl} Christoffel–symbol is the tensor of the gravitational field intensity. The dimension is m/s^2, acceleration.

The Λ "Lagrange–density" is actually the density of the drain (mass), that is the negative source density, thus some kind of vector divergence.

$\Lambda = div(\dot{g} \cdot a)$, where $\dot{g}$ is the gravitational acceleration and a is the graviton density.

Thus, the right side of the Einstein–equations is the spatial drain density, the left side is an expression related to the space–time curvature generated by the drain containing the field intensity (Christoffel–symbols).

gravitation spans the time–space continuum; this role is not played uniquely by gravitation.

The Einstein–equations are nonlinear, but for weak spaces including the classical Newtonian spaces, linearity gives a good approximation. What is important is to use the corresponding mechanisms to interpret the basic interactions.

We must conclude that there is no space–time in general, only gravitational space–time, electromagnetic space–time and strong– or weak space–time. Gravitational (or gravitomagnetic) space–time is governed point by point by the graviton field or, to be accurate, by the motion of the graviton field at that point. This constitutes the space–time structure of objects with rest–mass. Should there be particles only with electric charge but without (gravitating) rest–mass, gravitation would not have an effect on them and these objects would not create a gravitational space–time structure. (Such particles are not known, but there is no reason to exclude that they could exist.)

The space–time structure of the electric charge is governed by the photon field. If a particle has both charge and mass (which are the only kinds of particles we know), then both the photon field and the graviton field shape the structure of space–time.

The basic equations of superfluid–ether based grand unification

$$\delta(S_m + S_g + S_q + S_e) = \int \Big\{ \tfrac{c^3}{16G\pi} \Big[\big(R_{ik} - \tfrac{1}{2} g_{ik} R\big) \pm \tfrac{8G\pi}{c^4} {}^{g}T_{ik} \Big]$$

$$+ \tfrac{c^3}{16k\pi} \Big[\big(R_{ik} - \tfrac{1}{2} g_{ik} R\big) \pm \tfrac{8k\pi}{c^4} {}^{e}T_{ik} \Big] \Big\} g_{ik} \sqrt{-g}\, d\Omega$$

Where S_m is the effect of gravitating matter,

S_g is the effect of the gravitational field,

S_q is the effect of charge,

S_e is the effect of the electromagnetic field,

G is the gravitational constant,

$K = \frac{1}{4\pi_{\circ}}$, and $\varepsilon_{\circ}$ are the permittivity of vacuum,

${}^{g}T_{k}$ is the energy impulse tensor of the gravitational field, and ${}^{e}T_{k}$ is the energy impulse tensor of the electromagnetic field.

From this we get:

$$\delta(S_m + S_g + S_q + S_e) = \int\left\{\tfrac{c^3}{16G\pi}\left[(R_{ik} - \tfrac{1}{2} g_{ik} R) \pm \tfrac{8G\pi}{c^4}{}^{g}T_{ik}\right]\right.$$

$$\left. + \tfrac{c^3}{16k\pi}\left[(R_{ik} - \tfrac{1}{2} g_{ik} R) \pm \tfrac{8k\pi}{c^4}{}^{e}T_{ik}\right]\right\} g_{ik}\sqrt{-g}\,d\Omega$$

And from this we obtain:

$$\frac{R_k - \frac{1}{2} g_k R}{G} + \frac{R_k - \frac{1}{2} g_k R}{k} = \frac{8\pi}{c^4}\left(\pm^{g}T_k \pm^{e}T_k\right)$$

Multiplying by (G × k) from this we get in turn:

$$R_k - \tfrac{1}{2} g_k R(G + k) = \tfrac{8G\pi}{c^4}\left(\pm^{g}T_k \pm^{e}T_k\right)$$

At the level of atoms, the electric charges cancel out (for instance, the opposite charges of the electrons and protons within an atom). That is why physicists conclude that the structure of space–time is determined by gravitation only.

Applying the same logic to the strong/weak interactions, it is possible to unify the gravitational, electromagnetic and strong–weak fields into a single mathematical system. The 8equation shows how the three basic interactions determine the space–time structure of the particle within the nucleus at the given point.

The unification of the gravitational, electromagnetic and strong interactions

The space–time structure of the nucleons within the atom is not determined only by the gravitational and electromagnetic interactions, but to a far greater extent by the effect of the nuclear forces. Moreover, as there is a large order of magnitude difference between the nuclear forces and gravitation, the equation can be considered nothing more than a formalism. I will only write these equations because Einstein spent the second half of his life searching for them. From the age of 38 until his death at the age of 76 (in 1955) he devoted his life to this search but without success.

When writing these equations we immediately meet with a problem. The relativistic effects are strong, since the velocity of the nucleons in the nucleus is ν c/4. The spin dependence is also strong. (These we will disregard.)

$$R_k - \frac{1}{2} g_k R = \frac{Gk\ \alpha 8\pi}{c^4 (Gk + k\alpha + G\alpha)} (\pm^g T_k \pm^e T_k \pm^n T_k)$$

Where $\alpha = {}^n k e^{-r/r_0}$ and r_0 is the maximum atom radius. $^n k$ quantifies the difference in order of magnitude between the gravitational and nuclear forces. On the basis of the $^n k / G \quad 10^{39}$, quotient is $^n k$ 10^{28}

$$\frac{m}{kgs^2}$$

where $^n T_{ik}$ is the impulse tensor of the nuclear forces e^{-r/r_0} is the Yukawa distance dependence. r_0, the maximum atomic radius is essentially the radius of the action of the nuclear forces. This, according to experiments, is $r_0 = 1.1 \times 10^{-15}$ m.

The elements of $^n T_{ik}$ are the following:

$$^n T_{00} = \frac{p^2}{2m_n} \frac{A}{V}$$ is the energy density inside the atomic nucleus,

p = is the impulse of the particles,

m_n = is the mass of the nucleon.

The mass of the proton and neutron is identical up to the second decimal, $m_n = 1.67 \times 10^{-27}$ kg.

V = is the volume of the nucleus,

A = is the number of the nucleons.

We need to calculate using the maximum $^n T_{00}$, at which we arrive by using p maximum to substitute p.

$$^n T_{ik} (i \neq 0, k \neq 0) = \frac{d}{df^i} \frac{\max p_k}{t},$$

that is, the tensor of maximum tension. Here max p_k is the k dimension equivalent of the maximum impulse, t is the time and f^i is the surface.

$$^n T_{io} = {}^n T_{ok} (i \neq 0, k \neq 0) \approx c \cdot \max p_k \cdot \frac{A}{V}$$

is the maximum impulse density vector.

These are the basic equations of Grand Unification based on the Akasha viewed as a superfluid vacuum. The theory suggested by the equations has a formal aspect insofar as the structure of space–time is determined by different forces within the nucleus; beyond the nucleus; in molecular space; and finally in the macro–space between molecules.

Conclusions

The basic ideal of the Grand Unification of all the fields and forces of the universe can be re–attempted on the basis of the conception of the Akasha as a superfluid and superdense cosmic matrix. This matrix re–emerges as a reconceptualized version of the space–filling ether.

It appears that only a particular theory of the ether was disproven and subsequently discarded by the physics community at the beginning of the 20th century. The idea of the ether as a fundamental field or medium that underlies the manifest events of the universe has been neither disproven nor, as it now turns out, definitively discarded. Ever more theories at the cutting edge of contemporary physics indicate that the ether can be reconceptualized as a fundamental cosmic matrix, the Akasha that grounds the phenomena of the manifest world.

References

László Gazdag: Beyond the Theory of the Relativity. II, Robottechnika Kft., Budapest, 1998

László Gazdag: Superfluid mediums, vacuum spaces. Speculations in Science and Technology, Vol.12.,No. 1, 1989

László Gazdag: Einstein's second postulate. Speculations in Science and Technology, Vol. 18, 1995

László Gazdag: Combining of the strong and weak interactions. Light Work, USA, Tucson, 1998 február, Vol.4. No. 2a. p. 1–2.

László Gazdag: The Einstein–equations as real motion equations. Light Work, USA, Tucson, 1998. May.

László Gazdag: A teremtés titka. Alexandra Kiadó, Budapest, 2004.

Attila Grandpierre: The physics of collective consciousness. World Futures, 1997, Vol. 48.
L. D. Landau – E. M. Lifsic: Elméleti fizika II–III. (Theoretical physics) Tankönyvkiadó, Budapest, 1976, 1978
Sándor Máthé: Theoretical determination of the gravitational constant G. Galilean Electrodynamics, USA, Arlington, 1999. August, Vol.10, No.4.
Károly Simonyi: A fizika kultúrtörténete. (The culture–history of physics.) Gondolat Kiadó, Budapest, 1986

Annex III

The Transmuting Ether

Introduction to Subquantum Kinetics

Paul A. LaViolette

The Akasha is the "transmuting ether," an active cosmic matrix that gives rise to physical form. Its multifarious components, called "etherons," react among themselves, transforming and diffusing throughout space. Their interweaving processes bind the ether into organic unity.

Etherons in themselves have no mass, charge, or spin. The properties mass and spin appear when the etherons self-organize into soliton fields, etheric concentration patterns, which we recognize as material particles. Neutrons spontaneously nucleate from large magnitude electric and gravity fluctuations in the ether's vacuum state. Charge arises subsequently as a secondary organizational restructuring when a neutron progenitor spontaneously transforms into a positively charged proton with the emission of a negatively charged electron and antineutrino.

The ether is the ultimate reality; its concentration gradients are the prime cause of motion. Force is a derived stress-effect (a distortion of the subatomic particle's etheric soliton pattern) engendered when a concentration gradient is superimposed.

The new concept of the ether

Subquantum kinetics is a unified field theory whose description of microphysical phenomena has a general system theoretic foundation (LaViolette 1985a,b,c, 1994, 2010). It conceives subatomic particles as Turing wave patterns that self–organize within a subquantum medium functioning as an open reaction–diffusion system. This medium, termed the transmuting ether, is the Akasha. It forms the substrate from which all physical form emerge in the universe.

The transmuting ether, which requires more than three dimensions for its description, differs from 19th century mechanical ethers in that it is continually active, its multifarious components transmuting, reacting among themselves, and diffusing throughout space, with these interweaving processes binding the ether into an organic unity.

Subquantum kinetics presents a substantially different paradigm from that of standard physics which views particles as closed systems. Whether these be subatomic particles bound together by force fields, or quarks bound together by gluons, physics has traditionally conceived nature at its most basic level as composed of immutable structures. Unlike living systems which require a continuous flux of energy and matter connecting them with their environment in order to sustain their forms, conventional physics has viewed particles as self–sufficient entities, that require no interaction with their environment in order to continue their existence. Thus classical field theory leads to a conception of space that Alfred North Whitehead has criticized as being one of mere *simple location*, where objects simply have position without incorporating any reference to other regions of space and other durations of time.

Whitehead instead advocated a conception of space manifesting *prehensive unification*, where separate objects can be "together in space and together in time even if they be not contemporaneous." The ether (as the Akasha) of subquantum kinetics fulfills Whitehead's conception. As shown below, it is precisely because of its nonlinear, reactive, and interactive aspect that the transmuting ether of subquantum kinetics is able to spawn subatomic particles and photons, manifested either as stationary or inherently propagating ether concentration patterns. In the context of subquantum kinetics, the very existence of the physical world we see around us is evidence of the dynamic organic unity that operates in the universal substratum below, imperceptible to us and out of reach of direct detection by the most sophisticated instruments.

The notion of an ether, or of an absolute reference frame in space, necessarily conflicts with the postulate of special relativity that all frames are relative and the velocity of light is a universal constant. However, experiments by Sagnac (1913), Graneau (1983), Silvertooth (1987, 1989), Pappas and Vaughan (1990), Lafforgue (1991), and Cornille (1998), to name just a few, have established that the idea of relative frames is untenable and should be replaced with the notion of an absolute ether frame. Also a moderately simple experiment performed by Alexis Guy Obolensky has clocked speeds as high as 5 *c* for Coulomb shocks traveling across his laboratory (LaViolette, 2008a). Furthermore Podkletnov and Modanese (2011) report having measured a speed of 64 c for a collimated gravity impulse wave produced by a high voltage discharge emitted from a superconducting anode. These experiments not only soundly refute the special theory of relativity, but also indicate that information can be communicated at superluminal speeds.

However, subquantum kinetics does not negate the existence of "special relativistic effects" such as velocity dependent clock retardation and rod contraction. Nor, in offering an alternative to the space–time warping concept of general relativity, does it negate the reality of orbital precession, the bending of starlight, gravitational time dilation, and gravitational redshifting. These effects emerge as corollaries of its reaction–diffusion ether model (LaViolette 1985b, 1994, 2004, 2010).

It should be added that subquantum kinetics has made a number of testable predictions, twelve of which have been subsequently verified (LaViolette 1986, 1992, 1996, 2005, 2010), as listed in Table 1 on the next page.

The systems dynamics of subquantum kinetics

Subquantum kinetics was inspired by research on open chemical reaction systems such as the Belousov–Zhabotinskii (B–Z) reaction (Zaikin and Zhabotinskii 1970, Winfree 1974) and the Brusselator (Lefever 1968, Glansdorff and Prigogine 1971, Prigogine, Nicolis, and Babloyantz 1972, Nicolis and Prigogine 1977). Under the right conditions, the concentrations of the variable reactants of the

Table 1 Twelve Subsequently Verified *a priori* Predictions of Subquantum Kinetics

1 The prediction that the electric field in the core of a nucleon should be configured as a radially periodic Turing wave pattern of progressively declining amplitude, and that a charged nucleon should have a Turing wave pattern whose core electric potential is biased relative to the background electric potential (LaV, 1985b, 2008b).

2 The prediction that the universe is cosmologically stationary and that photons passing through intergalactic regions of space should progressively decrease their energy, that is, that photons should continually undergo a tired–light redshift effect (LaV, 1985c, 1986).

3 The prediction that photons traveling within galaxies should progressively increase their energy, that is, blueshift their wavelengths, and consequently that the luminosity of planets and red dwarf stars should be due to energy being spontaneously generated in their interiors (LaV, 1985c, 1992).

4 The prediction that the luminosity of brown dwarf stars should be due to the photon blueshifting effect described in (3) (LaV, 1985c, 1996, 2010).

5 The anticipation of the Pioneer effect; the prediction that a spacecraft maser signal transponded through interplanetary space should be observed to blueshift its wavelength at a rate of about one part in 10^{18} per second (LaV, 1985c, 2005).

6 The prediction that blue supergiant stars rather than red giant stars should be the precursors of supernova explosions (1985c, 1995).

7 The prediction that galactic core emissions should come from uncollapsed matter–creating stellar masses (Mother stars), rather than from matter–accreting black holes (LaV, 1985c).

8 The prediction that stars in the vicinity of the Galactic center should be massive blue supergiant stars as opposed to low mass red dwarf stars (LaV, 1985c).

9 The prediction that galaxies should progressively grow in size with the passage of time proceeding from compact types such as dwarf ellipticals and compact spirals to mature spirals and giant ellipticals (LaV, 1985c, 1994).

10 The prediction that a monopolar electron discharge should produce a longitudinal electric potential wave accompanied by a matter repelling gravity potential component (LaV, 1985b, 1994).

11 The prediction that the speed of the superluminal gravity wave component of a monopolar electron discharge should depend on the potential gradient of the discharge (LaV, 2003, 2010).

12 The prediction that inertial mass should decrease with a rise of G potential or with an increase in negative potential, and should increase with the reverse polarity (LaV, 1985b).

Brusselator reaction system can spontaneously self–organize into a stationary reaction–diffusion wave pattern such as that shown in figure 1. These have been called *Turing patterns* in recognition of Alan Turing who in 1952 was the first to point out their importance for biological morphogenesis. Alternatively, Prigogine et al. (1972) have referred to them as *dissipative structures* because the initial growth and subsequent maintenance of these patterns is due to the activity of the underlying energy–dissipating reaction processes. In addition, the B–Z reaction is found to exhibit propagating chemical concentration fronts, or *chemical waves* which may be easily reproduced in a school chemistry laboratory; see figure 2.

The Brusselator, the simpler of the two reaction systems, is defined by the following four kinetic equations:

$$
\begin{aligned}
A &\xrightarrow{k_1} X, && \text{a)} \\
B + X &\xrightarrow{k_2} Y + Z, && \text{b)} \\
2X + Y &\xrightarrow{k_3} 3X, && \text{c)} \\
X &\xrightarrow{k_4} \Omega. && \text{d)}
\end{aligned}
\qquad (1)
$$

The capital letters specify the concentrations of the various reaction species, and the k_i denote the kinetic constants for each reaction. Each reaction produces its products on the right at a rate equal to the product of the reactant concentrations on the left times its kinetic constant. Reaction species X and Y are allowed to vary in space and time, while A, B, Z and Ω are held constant.

This system defines two global reaction pathways which cross–couple to produce an X–Y reaction loop. One of the cross–coupling reactions, (1–c), is autocatalytic and prone to produce a nonlinear increase of X, which is kept in check by its complementary coupling reaction (1–b). Computer simulations of this system have shown that, when the reaction system operates in its supercritical mode, an initially homogeneous distribution of X and Y can self–organize into a wave pattern of well–defined wavelength in which X and Y vary reciprocally with respect to one another as shown in Figure 1. In other words, these systems allow order to spontaneously emerge (entropy to decrease) by virtue of the fact that they function as open

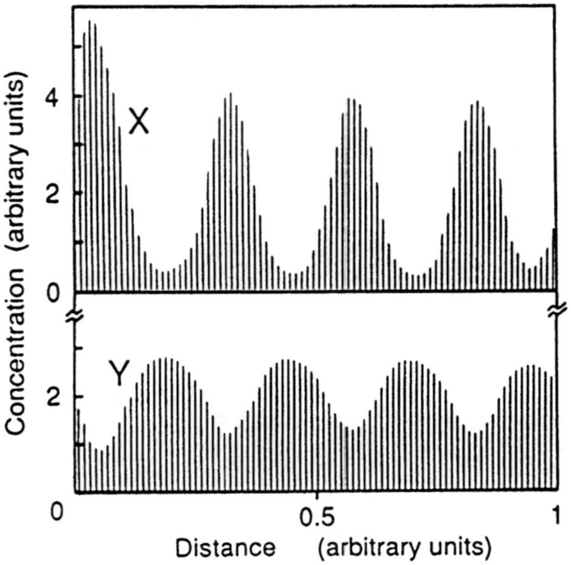

Figure 1. One-dimensional computer simulation of the concentrations of the Brusselator's X and Y variables (diagram after R. Lefever 1968).

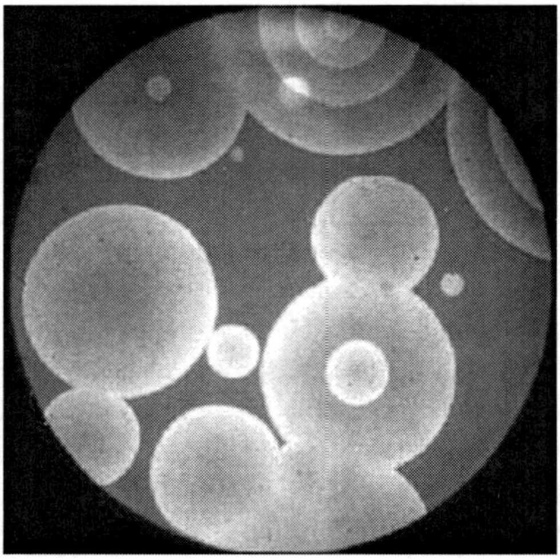

Figure 2. Chemical waves formed by the Belousov–Zhabotinskii reaction (photo courtesy of A. Winfree).

systems; the Second Law of Thermodynamics only applying to closed systems.

The Model G ether reaction system

Subquantum kinetics postulates a nonlinear reaction system similar to the Brusselator that involves the following five kinetic equations termed *Model G* (LaViolette, 1985b):

$$A \underset{k_{-1}}{\overset{k_1}{\rightleftarrows}} G, \qquad \text{a)}$$

$$G \underset{k_{-2}}{\overset{k_2}{\rightleftarrows}} X, \qquad \text{b)}$$

$$B + X \underset{k_{-3}}{\overset{k_3}{\rightleftarrows}} Y + Z, \qquad \text{c)} \qquad (2)$$

$$2X + Y \underset{k_{-4}}{\overset{k_4}{\rightleftarrows}} 3X, \qquad \text{d)}$$

$$X \underset{k_{-5}}{\overset{k_5}{\rightleftarrows}} \Omega. \qquad \text{e)}$$

The kinetic constants k_i denote the relative propensity for the reaction to proceed forward, and k_{-i} denote the relative propensity for the corresponding reaction to proceed in the reverse direction. The forwarded reactions are mapped out in figure 3. Since the forward kinetic constants have values much greater than the reverse kinetic constants, the reactions have the overall tendency to proceed irreversibly to the right. Nevertheless, the reverse reactions, in particular that associated with reaction (2–b), play an important role. Not only does this one allow Model G to establish electro–gravitic field coupling, but as described below, it also allows the spontaneous formation of material particles in an initially subcritical ether.

Whereas the Brusselator and B–Z reaction conceive of a chemical medium consisting of various reacting and diffusing molecular species, subquantum kinetics conceives of a space–filling etheric medium consisting of various reacting and diffusing *etheric* species termed etherons. Being present as various etheron types (or states) labeled A, B, X, and so on, these diffuse through space and react with one another in the manner specified by Model G. (Figure 3) Model G is in effect the recipe, or software, that

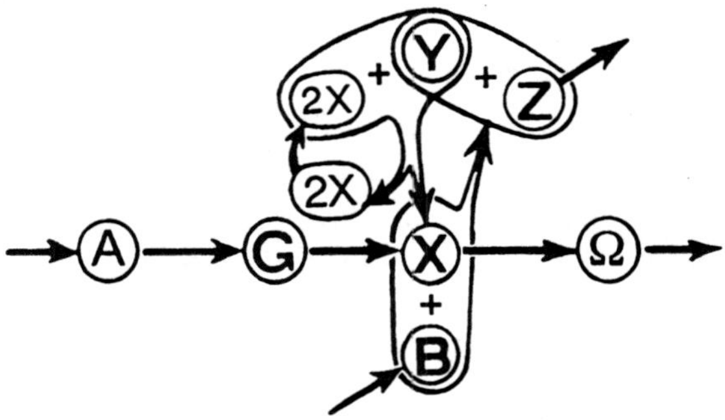

Figure 3. The Model G ether reaction system investigated by subquantum kinetics.

generates the physical universe. Etherons should not be confused with quarks. Whereas quark theory proposes that quarks exist only within the nucleon, with just three residing within each such particle, subquantum kinetics presumes that etherons are far more ubiquitous, residing not only within the nucleon, but also filling all space with a number density of over 10^{25} per cubic fermi, where they serve as the substrate for all particles and fields.

The self closing character of the X–Y reaction loop, which is readily evident in figure 3, is what allows Model G and the Brusselator to generate ordered wave patterns. Model G is similar to the Brusselator with the exception that a third intermediary variable, G, is added with the result that steps (2–a) and (2–b) now replace step (1–a) of the Brusselator, all other steps remaining the same. The third variable was introduced in order to give the system the ability to nucleate self–stabilizing localized Turing patterns within a prevailing subcritical environment. This *autogenic* particle formation ability is what allows Model G to become a promising candidate system for the generation of physically realistic subatomic structures.

Based on the reaction equation system (3) we may write the following set of partial differential equations to depict how all three reaction intermediates G, X and Y vary as a function of space and time in three dimensions:

$$\left.\begin{aligned}
\frac{\partial G(x, y, z, t)}{\partial t} &= k_1A - k_2G + D_g\nabla^2G \\
\frac{\partial X(x, y, z, t)}{\partial t} &= k_2G + k_4X^2Y - k_3BX - k_5X + D_x\nabla^2X \\
\frac{\partial Y(x, y, z, t)}{\partial t} &= k_3BX - k_4X^2Y + D_y\nabla^2Y
\end{aligned}\right\} \quad (3)$$

where the D_g, D_X and D_y values represent the diffusion coefficients of the respective reactant variables.

A homogeneous distribution of the G, X, and Y reaction intermediates would correspond to a spatial vacuum devoid of matter and energy. Variations in the concentrations of these three variables would correspond to the formation of observable electric and gravitational potential fields, and wave patterns formed by these fields, in turn, would constitute observable material particles and energy waves. The etherons themselves would remain unobservable. Subquantum kinetics identifies the G concentration with gravitational potential, where G concentrations greater than the prevailing homogeneous steady–state concentration value, G_O, would constitute positive gravity potentials and G concentrations less than G_O would constitute negative gravity potentials. A negative G potential well, G ether concentration well, would correspond to a matter–attracting negative gravity potential field, whereas a positive G potential hill would correspond to a matter–repelling positive gravity potential field. The X and Y concentrations, which are mutually interrelated in reciprocal fashion, are together identified with electric potential fields. A configuration in which the Y concentration is greater than Y_O and the X concentration is less than X_O would correspond to a positive electric potential and the opposite polarity, low–Y/high–X would correspond to a negative electric potential. Relative motion of an electric potential field, or of an X–Y concentration gradient, would generate a magnetic (or electrodynamic) force (LaViolette 1994, 2010). As Feynman, Leighton, and Sands (1964) have shown, in standard physics magnetic force can be mathematically expressed solely in terms of the effect that a moving electric potential field produces on a charged particle obviating the need for magnetic potential field terms. Also relative motion of a gravity potential field, of a G concentration gradient, is predicted to generate a gravitodynamic force, the gravitational equivalent of a magnetic force.

The subquantum kinetics ether functions as an open system, where etherons transform irreversibly through a series of "upstream" states, including states A and B, eventually occupying states G, X, and Y, and subsequently transforming into the D and Ω states and from there through a sequence of "downstream" states; see figure 4. This irreversible sequential transformation is conceived as defining a vectorial dimension line termed the transformation dimension. Our observable physical universe would be entirely encompassed by the G, X, and Y ether states, which would reside at a nexus along this transformation dimension, the continual etheron transformation process serving as the Prime Mover of our universe. According to subquantum kinetics, the arrow of time, as physically observed in all temporal events, may be attributed to the continuation of this subquantum transformative process. Subquantum kinetics allows the possibility of parallel universes forming either "upstream" or "downstream" of our own universe wherever the ether reaction stream intersects itself to form a reaction loop similar to Model G. However, while there is a finite chance of such a material universe being spawned, the possibility that it would actually form are vanishingly small since the ether reaction parameters would need to adopt the proper precise values in order to spawn the necessary nucleon building blocks.

Since etherons both enter and leave the etheron states that compose material bodies and energy waves in our physical universe,

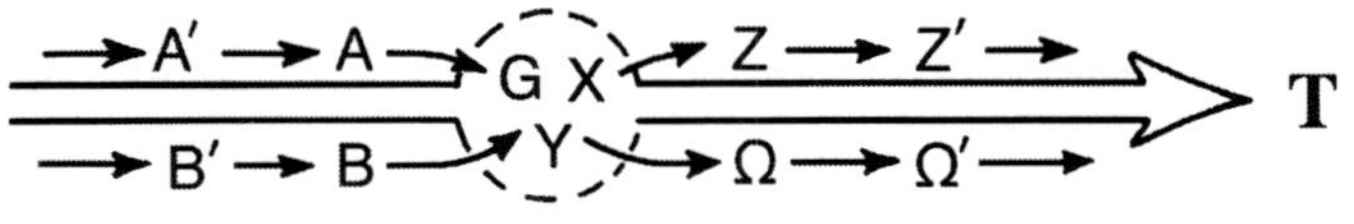

Figure 4. An expansion of the Model–G ether reaction scheme as it would appear disposed along dimension **T**. G, X, and Y mark the domain of the physical universe.

our observable universe may be said to be open to the throughput of etherons. That is, our universe would function as an *open system*. In such a system, ordered field patterns may spontaneously emerge from initially homogeneous field distributions or they may progressively dissolve back to the homogeneous state depending on the criticality of the reaction system.

In Model G, the system's criticality is determined by the value of the G variable. Sufficiently negative G potentials create supercritical conditions that allow matter formation and photon blueshifting while positive G potential values create subcritical conditions that cause tired-light photon redshifting and in extreme instances particle dematerialization.

The transmuting ether of subquantum kinetics bears some resemblance to the ether concept of Nikola Tesla. He proposed a gas-like ether that is acted on by a "life-giving creative force" which when thrown into infinitesimal whorls gives rise to ponderable matter and that when this force subsides and the motion ceases, matter disappears leaving only the ether. In subquantum kinetics, this creative force or Prime Mover is termed *etheric Force* while the resulting transmutative or reactive transformation of etherons from one state to another is termed *etheric flux*.

The transmuting ether also closely parallels the descriptions of Besant and Leadbeater (1919) who as early as 1895 said "the ether is not homogeneous but consists of particles of numerous kinds, differing in the aggregations of the minute bodies composing them." About the subatomic particle, which they refer to as the "ultimate physical atom", they state: "It is formed by the flow of the life–force and vanishes with its ebb. When this force arises in 'space'... [u.p.a] atoms appear; if this be artificially stopped for a single atom, the atom disappears; there is nothing left. Presumably, were that flow checked but for an instant, the whole physical world would vanish, as a cloud melts away in the empyrean. It is only the persistence of that flow which maintains the physical basis of the universe." Similarly, subquantum kinetics views our observable physical universe as an epiphenomenal watermark generated by the activity of a higher dimensional ether that functions as an open system.

Parthenogenesis: The creation of matter from zero–point fluctuations

According to subquantum kinetics, material particles nucleate from electric and gravity potential fluctuations that spontaneously arise from the ether vacuum state. Since etherons react and transform in a stochastic Markovian fashion, the etheron concentrations of all etheron species will vary stochastically above and below their steady–state values, the magnitudes of these fluctuations conforming to a Poisson distribution. It is known that such fluctuations are present in the chemical species of reaction–diffusion systems such as the B–Z reaction and their presence is also postulated in the theoretical Brusselator system. So the same would be true in the Model G reactive ether. Hence subquantum kinetics predicts that stochastic electric and gravity potential fluctuations should spontaneously arise throughout all of space, in regions both where field gradients are present and where they are absent.

This is in some ways analogous to the concept of the zero–point energy (ZPE) background, but with some differences. In conventional physics, ZPE fluctuations are theorized to have energies comparable to the rest mass energy of subatomic particle and to emerge as particle–antiparticle pairs which rapidly annihilate one another. As a result, it is fashionable to quote unimaginably high values of the order of 10^{36} to 10^{113} ergs/cm^3 for the zero–point energy density of space. But because of their polarity pairing, they are unable to nucleate matter. By comparison, subquantum kinetics rejects the idea that the spatial vacuum is "seething with particles and antiparticles." It theorizes far lower ZPE densities, on the order of less than 1 erg/cm^3, or less than the radiation energy density prevailing at 2000° K. Nevertheless, because these fluctuations are unpaired, they are potentially able to spawn material particles. But this only occurs when a fluctuation of sufficiently large magnitude arises, the vast majority being far too small to attain the required subquantum energy threshold.

Provided that the kinetic constants and diffusion coefficients of the ether reactions are properly specified to render the system subcritical but close to the critical threshold, a sufficiently large spontaneously arising positive zero–point electric potential fluctuation (i.e., a critical fluctuation consisting of a low X concentration or a high Y concentration), with further growth, a further reduction of X and increase in Y, is able to break the symmetry of the initial vacuum state to produce what is called a *Turing bifurcation.* That is, it is able to change the initially uniform electric and gravity potential background field that defines the vacuum state into a localized, steady–state periodic structure. In subquantum kinetics this emergent wave pattern would form the central electric and gravity field structure of a nascent subatomic particle.

One advantage of Model G is that a positive electric potential fluctuation characterized by a negative X potential also generates a corresponding negative G potential fluctuation by virtue of the reverse reaction $X \xleftarrow{k_{-2}} G$, and this in turn produces a local supercritical region allowing the seed fluctuation to persist and grow in size. Consequently, if the ether reaction system is initially in the subcritical vacuum state, provided that it operates sufficiently close to the critical threshold, eventually a fluctuation will arise that is sufficiently large to form a supercritical region and nucleate a subatomic particle (e.g., a neutron). Thus spontaneous matter and energy creation is allowed in subquantum kinetics.

This parthenogenic, order-through-fluctuation process is shown in figure 5, which presents successive frames from a 3D computer simulation of equation system (3) (Pulver and LaViolette, 2011). Spherical symmetry was imposed as an arbitrary assumption to reduce the computing time necessary to carry out the simulation. The duration of the simulation consists of 100 arbitrary time units and the reaction volume measures 100 arbitrary spatial units, from –50 to +50, with one fifth of the volume being displayed in the graph. Vacuum boundary conditions are assumed. These space and time units are dimensionless, meaning that the units of measure are not specified. To initiate the particle's nucleation, a negative *subquantum* X ether fluctuation $-\varphi_x(r)$ was introduced at spatial coordinate $r = 0$. The rise and fall of the fluctuation magnitude reaches its maximum value of –1 after 10 time units, or 10% of the way through the simulation, and diminishes back to zero magnitude (flat line) by 20 units of time, or 20% of the way through the simulation. The reaction system quickly generates

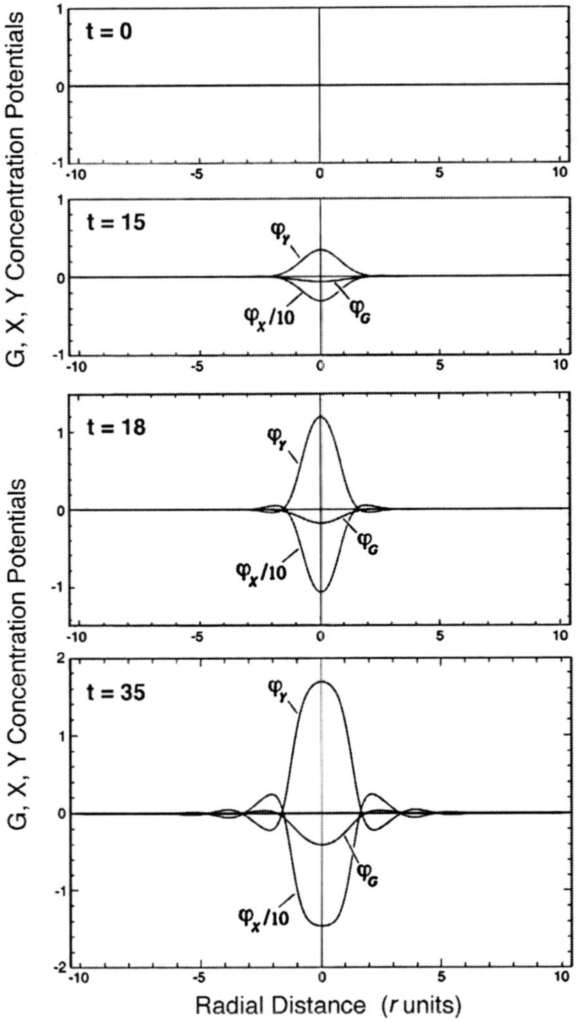

Figure 5. Sequential frames from a three–dimensional computer simulation of Model G showing the emergence of an autonomous dissipative structure particle: t = 0 the initial steady state; t = 15 growth of the positively charged core as the X seed fluctuation fades; t = 18 deployment of the periodic electric field Turing wave pattern; and t = 35 the mature dissipative structure particle maintaining its own supercritical core G–well. Simulation by M. Pulver.

a complementary positive Y potential fluctuation $+\varphi_y(r)$, which together comprise a positive electric potential fluctuation, and also generates a negative G fluctuation $-\varphi_g(r)$, which comprises a gravity potential well. This is apparent in the second frame at t = 15 units. This central G-well generates a region that is sufficiently supercritical to allow the fluctuation to rapidly grow in size and eventually develop into an autonomous particulate dissipative structure which is seen fully developed in the last frame at t = 35 units. Movies of this and other Model G simulations are posted at www. starburstfound.org/simulations/archive.html.

The particle shown here would represent a neutron. It's electric field consists of a Gaussian central core of high-Y/low-X polarity surrounded by a pattern of concentric spherical shells where X and Y alternate between high and low extrema of progressively declining amplitude. Being a reaction-diffusion wave pattern, we may appropriately name this periodicity the particle's *Turing wave* (LaViolette, 2008b). The antineutron would have the opposite polarity, high-X/low-Y centered on a G potential hill.

The positive Y potential field (negative X potential field) in the neutron's core corresponds to the existence of a positive electric charge density, and the surrounding shell pattern which alternates between low and high Y potentials constitutes shells of alternating negative and positive charge density. On the average, however, these charge densities cancel out to zero in the case of the neutron, which is why the Turing wave for the simulated neutron shown in figure 5 has no positive or negative bias with respect to the ambient zero potential.

The appearance of these positive and negative charge densities necessitates the simultaneous appearance of the particle's inertial rest mass. The shorter the wavelength of the Turing wave, and greater its amplitude (greater its etheron concentration wave amplitude), the greater will be the inertial mass of the associated particle (LaViolette 1985b). Since acceleration requires a structural shift and recreation of the particle's Turing-wave dissipative space structure, the particle's resistance to acceleration, its inertia, should be proportional to the magnitude of its Turing-wave charge densities; that is, proportional to the amount of negentropy that must be restructured (LaViolette 2010).

Subquantum kinetics further requires that for Model G to be physically realistic, the values of its kinetic constants, diffusion coefficients, and reactant concentrations should be chosen such that the emergent Turing wave has a wavelength equal to the Compton wavelength, λ_0, of the particle it represents, this being related to particle rest mass energy E_o, or to its rest mass m_o, by the formula:

$$\lambda_0 = hc/E_o = h/m_o c, \qquad (4)$$

where h is Planck's constant and c is the velocity of light. The Compton wavelength for the nucleon calculates to be 1.32 fermis ($\lambda_0 = 1.32 \times 10^{-13}$ cm). This prediction that a particle's core electric field should have a Compton wavelength periodicity has since been confirmed by particle scattering experiments; see below. Moreover, unlike the Schrödinger linear wave packet representation of the particle, which has the unfortunate tendency to progressively dissipate over time, the localized dissipative structures predicted by Model G maintain their coherence, the underlying ether reaction-diffusion processes continuously combat the increase of entropy. Thus the Schrödinger wave equation used in quantum mechanics offers a rather naive linear approximation to representing microphysical phenomena, the quantum level being better described by a nonlinear equation system such as Model G.

Since this Turing wave particle representation incorporates both particle and wave aspects, we are able to dispense with the need to adopt a wave-particle dualism view of quantum interactions. Moreover the Turing wave subatomic particle has been shown to quantitatively account for the results of particle diffraction experiments, thereby eliminating paradoxes that arise in standard theories that rely on deBroglie's pilot wave interpretation or Schrödinger's wave packet concept. It also correctly yields Bohr's orbital quantization formula for the hydrogen atom while at the same time predicting a particle wavelength for the ground state orbital electron that is ~1400 times smaller than Schrödinger's wave packet prediction (LaViolette, 1985, 2008b, 2010). This more compact representation of the electron, allows the existence of smaller diameter sub-ground state orbits having fractional quantum numbers. Several researchers, such as John Eccles and Randall Mills, claim to have developed methods of inducing electron transitions to such subground orbits and to thereby extract enormous quantities of energy from plain water (LaViolette, 2008b). So reformulating quantum mechanics on the basis of the subquantum kinetics Turing wave concept, opens the door to understanding and developing new environmentally safe technologies that could power our world.

In the course of dispensing with the Schrödinger wave packet and its associated probability function describing the indeterminate position of a mass point, it is advisable to also throw out the Copenhagen interpretation with its mysterious "collapse of the wave function" theorized to take place when the quantum "entity" through measurement becomes determined to be either a wave or a particle. In particular, Dewdney et al. (1985) have shown experimentally that the position of the particle is defined in a real sense prior to its deBroglie scattering event and from this conclude that in this particular case the wave-packet-collapse concept is flawed. More than likely, we should also be able to avoid using this collapse concept also in experiments observing the spin orientation of entangled particles or polarization of entangled photons. There appears to be the growing realization that its widespread use is mainly a convenient mechanism to cover up the fact that we currently have a poor understanding of the workings of the subquantum realm.

When a neutron spontaneously acquires *positive charge* and transforms into a proton, its X–Y wave pattern acquires a positive bias similar to that shown in Figure 6 (shaded region in the left hand plot). Such a biasing phenomenon, which is seen in analysis of the Brusselator, is also present in Model G when an existing ordered state undergoes a *secondary bifurcation*. The transition of the neutron to the positively biased proton state is best understood by reference to a bifurcation diagram similar to those used to represent the appearance of ordered states in nonequilibrium chemical reaction systems (see Figure 7).

The emergence of the neutron from the vacuum state is represented as a transition to the upper primary bifurcation branch which branches past critical threshold β_C. Past threshold β' this

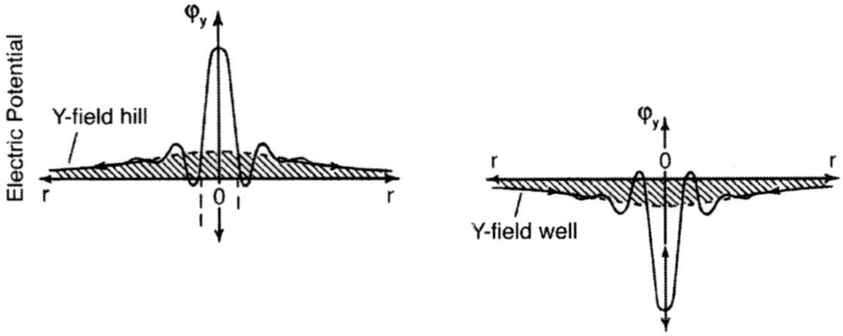

Figure 6. Radial electrostatic potential profiles for a proton and antiproton, positively charged matter state (left) and negatively charged antimatter state (right). The characteristic wavelength equals the particle's Compton wavelength.

primary branch undergoes a secondary bifurcation with the emergence of the proton solution branch. This transition is observed in the phenomenon of beta decay, which also involves the production of electron and antineutrino particles not charted here; i.e., $n \rightarrow p + e^- + \overline{\nu}^o + \gamma$.

The neutron's transition to the charged proton state involves an excess production rate of Y per unit volume in its core coupled with a corresponding excess consumption rate of X per unit

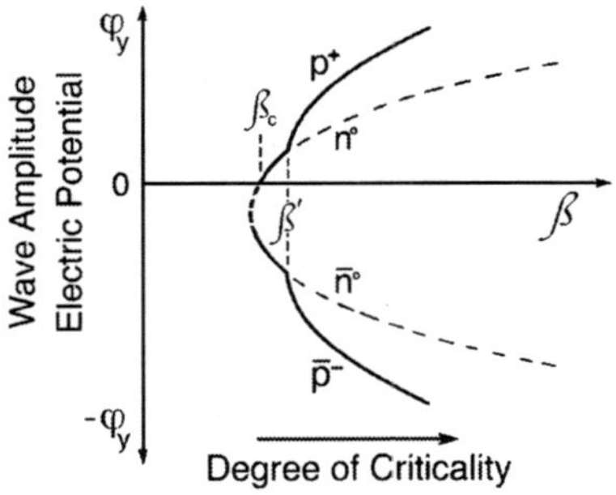

Figure 7. A hypothetical bifurcation diagram for the formation of nuclear particles. the secondary bifurcation past bifurcation point β' creates electrostatic charge.

volume. This causes a positive bias in its central Y concentration and a negative bias in its central X concentration, which in turn extends radially outward to bias the particle's entire Turing wave pattern. This extended field bias constitutes the particles long-range electric field. Analysis shows that this potential bias declines as the inverse of radial distance just as classical theory predicts. In fact, subquantum kinetics has been shown to reproduce all the classical laws of electrostatics, as well as all the classical laws of gravitation (LaViolette, 1985b, 1994, 2010).

It should be kept in mind that the charge densities forming the proton's Turing wave pattern, and that are associated with its inertial mass, are distinct from and additional to the charge density that centrally biases its Turing pattern and produces the particle's long-range electric field. The former periodic densities emerge as a result of the particle's primary bifurcation from the homogeneous steady-state solution, while the latter aperiodic bias emerges as a result of its secondary bifurcation from an existing steady-state Turing solution.

Based on the results of the Sherwin-Rawcliffe experiment, we may infer that the creation and later displacement of the particle's Turing wave field would be communicated outward essentially instantaneously or at an exceedingly high superluminal velocity. The same would hold for the outward moving event boundaries of a subatomic particle's long-range electric and gravity potential fields. For their experiment, Sherwin and Rawcliffe (1960) performed mass spectrometry measurements of a football-shaped Lu^{175} nucleus to check for the presence of line-splitting and came up with a null result. This indicated that the mass of the lutetium nucleus behaved as a scalar instead of a tensor which implied that its Coulomb field moved rigidly with its nucleus and was thereby able to create instant action-at-a-distance (Phipps, 2009). Accordingly, the conventional practice of retarding force actions at speed c would be inappropriate.

Subquantum kinetics leads to a novel understanding of force, acceleration, and motion. In subquantum kinetics the energy potential field (ether concentration gradient) is regarded as the real existent and the prime cause of motion, "force" being regarded as a *derived* manifestation. That is, force is interpreted as the *stress effect* which the potential gradient produces on the material particle due to the distortion it manifests on the field pattern space structure that composes the particle. The particle relieves this stress by homeostatic adjustment which results in a jump acceleration and relative motion.

Particle scattering confirmation

The Turing wave configuration of the nucleon's electric potential field predicted by subquantum kinetics has been confirmed by particle scattering experiments that employ the recoil-polarization technique. Kelly (2002) has obtained a good fit to particle scattering form factor data by representing the radial variation of charge and magnetization density with a relativistic Laguerre-Gaussian expansion; see figures 8-a and 9-a. The periodic character of this fit is more apparent when surface charge density ($r^2\rho$) is plotted as a function of radial distance as shown in figures 8-b and 9-b. Kelly's charge density model predicts that the proton and neutron both have a gausian shaped positive charge density core surrounded by a periodic electric field having a wavelength approximating the Compton wavelength. Moreover he has noted that unless this surrounding periodicity is included, his nucleon charge and magnetization density models do not make as good a fit to form factor data.

Thus here we have a stunning confirmation of a central feature of the subquantum kinetics physics methodology whose prediction was first made in the mid 70's at a time when it was still convention to regard the field in the core of the nucleon as rising sharply to a central cusp. Note also that Kelly's model confirms the positive biasing of the proton's central field, the bias increasing as the center of the particle is approached; compare the enhanced view shown in figure 9-b with figure 6. Furthermore, as in the subquantum kinetics model, Kelly's model shows the amplitude of the nucleon's peripheral periodicity declining with increasing radial distance.

Simulations performed on Model G show that the amplitude of the Turing wave pattern declines with increasing radial distance

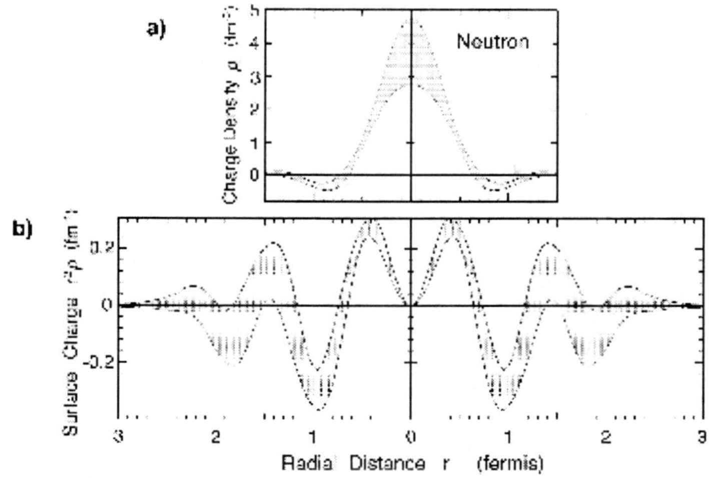

Figure 8. a) Charge density profile for the neutron predicted by Kelly's preferred Laguerre-Gaussian expansion models and b) the corresponding surface charge density profile (after Kelly 2002, Fig. 5-7, 18).

as $1/r^4$ at small radii ($r < 2\lambda_0$), which approximates the radial decline observed in the charge density maxima for Kelly's model. The Model G particle Turing wave pattern declines more steeply at greater radial distances, declining as $1/r^7$ at $r \approx 4\lambda_0$ and $1/r^{10}$ for $r \approx 6\lambda_0$, which may be compared to standard theory which propose that the nuclear force declines as F_n μ $1/r^7$. This localized particle wave pattern is possible only because the extra G variable has been introduced into the Model G reaction system. It allows a particle to self-nucleate in an initially subcritical environment while leaving distant regions of space in their subcritical vacuum state. Thus if we quantify the amount of order or negentropy created by a single seed fluctuation, integrate the total amount of field potential $|\varphi_x|$ or $|\varphi_y|$ forming the particle wave pattern, we should find that it converges to a finite value, comparable to the idea of a quantum of action. The two-variable Brusselator, on the other hand, fails to generate localized structures. Simulations show that a seed fluctuation in the Brusselator produces order only if the

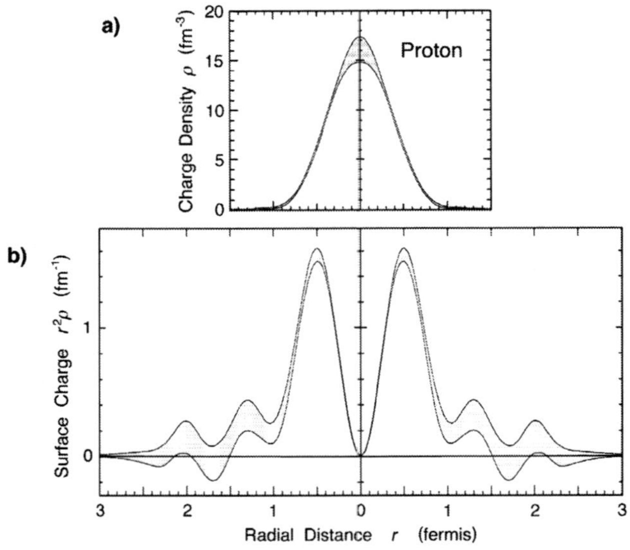

Figure 9. a) Charge density profile for the proton predicted by Kelly's preferred Laguerre-Gaussian expansion models and b) the corresponding surface charge density profile (after Kelly 2002, Fig. 5–7, 18).

system initially operates in the supercritical state, which in turn causes its entire reaction volume to become filled with a Turing wave pattern of maximum amplitude. Thus in the Brusselator a single seed fluctuation potentially produces an infinite amount of negentropy or structure.

The confirmation of the Model G ether reaction model which has been forthcoming from particle scattering experiment data leads us to conceive of the subatomic particle as an organized entity, or system, whose form is created through the active interplay of a plurality of particulate structures existing at a lower hierarchic level. Thus we find that the very structure of matter, its observationally confirmed Turing wave character, stands as proof of an underlying Whiteheadian dynamic and interactive stratum, one that ancient cultures variously named the Aether, Akasha, Tao, or Cosmic Ocean. The physics of subquantum kinetics has very ancient origins indeed (LaViolette, 2004).

Contemporary quark models fail to anticipate the periodic character of the nucleon's electric field. No quark model can be devised after-the-fact that can reasonably account for this feature. Quarks themselves, or the "gluons" theorized to bind them together, have no script to tell them they should dance around in the complex manner that would be required in order to generate such an extended periodic field pattern. Subquantum kinetics, the viable replacement for quark theory, differs in several respects, one being the manner in which it handles the origin of mass, charge and spin. Quark theory does not attempt to explain how inertial mass, electric charge or spin arise. It merely assumes them to be physical attributes present in quarks in fractional form and which in triplicate summation appear as corresponding properties detectable in the nucleon. By comparison, etheron reactants of subquantum kinetics have no mass, charge, or spin. These are properties which are predicted to arise only at the quantum level, and which amazingly emerge as corollaries of the Model G reactions. Mass and spin, as properties of the subatomic particle, emerge at the time the particle first comes into being, and charge, as noted earlier, emerges as a secondary bifurcation of the primary Turing bifurcation.

References

AUCHMUTY, J. F. G. and NICOLIS, G., 1975, Bifurcation analysis of nonlinear reaction diffusion equations—1. Evolution equations and the steady state solutions. *Bulletin Mathematical Biology.* **37** pp. 323–365.

BESANT, A. and LEADBEATER, C. W., 1919, *Occult Chemistry: Clairvoyant Observations on the Chemical Elements.* London: Theosophical Publishing House.

BRUMFIEL, G., 2008, Physicists spooked by faster-than-light information transfer. *Nature*: 1038. Naturenews: http://www.nature.com/news/2008/080813/full/news.2008. 1038.html.

CORNILLE, P., 1998, Making a Trouton-Noble experiment succeed. *Gal. Electrod.* **9** (1998): 33.

DEWDNEY, C. et al., 1985, *Foundations of Physics* **15**, pp. 1031–1042.

FEYNMAN, R. P., LEIGHTON, R. B., and SANDS, M. 1964, *The Feynman Lectures on Physics*, Vol. II, Reading MA: Addison-Wesley.

GLANSDORFF, P. and PRIGOGINE., I., 1971, *Thermodynamic Theory of Structure, Stability, and Fluctuation*. New York: Wiley.

GRANEAU, N., 1983, First indication of Ampere tension in solid electric conductors. *Phys. Lett.* **97A**: 253–255.

KELLY, J., 2002, Nucleon charge and magnetization densities from Sachs form factors. *Physical Review C* **66** (6), id: 065203. Eprint: http://arXiv.org/abs/hep-ph/0204239.

LAFFORGUE, J.-C., 1991, Isolated systems self-propelled by electrostatic forces. French patent No. 2651388.

LAVIOLETTE, P. A., 1985a, An introduction to subquantum kinetics. I. An overview of the methodology. *Intern. J. General Systems* **11**, pp. 281 – 293.

LAVIOLETTE, P. A., 1985b, An introduction to subquantum kinetics: II. An open systems description of particles and fields. *Intern. J. General Systems* **11**, pp. 305–328.

LAVIOLETTE, P. A., 1985c, An introduction to subquantum kinetics: III. The cosmology of subquantum kinetics. *Intern. J. General Systems* **11**, pp. 329–345.

LAVIOLETTE, P. A., 1986, "Is the universe really expanding?" *The Astrophysical Journal* **301**, pp. 544–553.

LAVIOLETTE, P. A., 1992, "The Planetary-stellar mass-luminosity relation: Possible evidence of energy nonconservation?" *Physics Essays* **5**(4), pp. 536–544.

LAVIOLETTE, P. A., 1994, *Subquantum Kinetics: The Alchemy of Creation*. Alexandria, VA: Starlane Publications, first edition (out of print).

LAVIOLETTE, P. A., 1996, "Brown dwarf discovery confirms theory of spontaneous energy generation." *Infinite Energy* **1** (5 & 6): 31; reprinted in *New Energy News* **3** (9) (1996): 17–18.

LAVIOLETTE, P. A., 2004, *Genesis of the Cosmos* (Rochester, VT, Bear & Co.), first ed. 1995.

LAVIOLETTE, P. A., 2005, "The Pioneer maser signal anomaly: Possible confirmation of spontaneous photon blueshifting." *Physics Essays* **18**(2), pp. 150–163.

LAVIOLETTE, P. A., 2008a, *Secrets of Antigravity Propulsion*. Rochester, VT: Bear & Co.

LAVIOLETTE, P. A., 2008b *International Journal of General Systems*, vol. 37 (6) (2008): 649–676.

LAVIOLETTE, P. A., 2010, *Subquantum Kinetics: A Systems Approach to Physics and Astronomy*. Niskayuna, NY: Starlane Publications, third edition, second ed. 2003.

LEFEVER, R., 1968, Dissipative structures in chemical systems. *J. Chemical Physics* **49**, pp. 4977–4978.

NICOLIS, G. and PRIGOGINE, I., 1977, *Self-organization in Nonequilibrium Systems*. New York: Wiley-Interscience.

PAPPAS, P. T. and VAUGHAN, T., 1990, Forces on a stigma antenna." *Physics Essays* **3**: 211–216.

PHIPPS, Jr., T. E., 2009, The Sherwin-Rawcliffe experiment - Evidence for instant action-at-a-distance. *Apeiron* **16**: 503–515.

PODKLETNOV, E. and MODENESE, G., 2011, Study of light interaction with gravity impulses and measurements of the speed of gravity impulses. In: *Gravity-Superconductors Interactions: Theory and Experiment*, edited by G. Modanese and R. Robertson. Bussum, Netherlands: Bentham Science Publishers.

PRIGOGINE, I., NICOLIS, G. and BABLOYANTZ, A., 1972, Thermodynamics of Evolution. *Physics Today* **25**(11), pp. 23–28; **25** (12), pp. 38–44.

PULVER, M. and LAVIOLETTE, P. A., 2011, Stationary particle formation in a reaction-diffusion system. paper in preparation.

SAGNAC, G. 1913, The luminiferous ether demonstrated by the effect of the relative motion of the ether in an interferometer in uniform rotation." *Comptes Rendus de l'Academie des Sciences* (Paris) **157**: 708–710.

SHERWIN, C. W. and RAWCLIFFE, R. D., *Report I-92 of March 14, 1960 of the Consolidated Science Laboratory* (University of Illinois, Urbana); Dept. of Commerce Scientific and Technical Information Service, document #625706.

SILVERTOOTH, E. W., 1987, Experimental detection of the ether." *Speculations in Science and Technology* **10**: 3–7.

SILVERTOOTH, E. W., 1989, Motion through the ether." *Electronics and Wireless World*: 437–438.

WINFREE, A. T., 1974 , Rotating chemical reactions. *Scientific American* **230**, pp. 82–95.

ZAIKIN, A. and ZHABOTINSKII, A., 1970, Concentration wave propagation in two-dimensional liquid-phase self-oscillating system. *Nature*, **225**, pp. 535–537.

Annex IV

The Aether Model

A New Foundation for Physics

David W. Thomson III
and James D. Bourassa

The Aether Model (also known as the Aether Physics Model, APM) unifies the four universal fields and forces of nature (the electromagnetic, the gravitational, and the strong and the weak interactions) by a set of laws rooted in the fabric of spacetime. The Akasha—called "Aether"—is a fabric of rotating magnetic fields with an electromagnetic, electrostatic, and gravitational dipole structure. Subatomic particles are viewed as angular momentum encapsulated in a rotating magnetic quantum field. All quantum, atomic, and molecular processes can be derived as originating in a primal "singularity" that is complete and whole, without dualities, space, time, and manifest matter.

Introduction

The idea that the observable world derives in some ultimate sense from a deeper level of reality did not cease to fascinate poets, prophets, and philosophers. Today, the quantum level of reality fascinates physical scientists. It is not surprising, then, that at the leading edge of the physical sciences a new paradigm should be taking shape, based on the quantum vacuum—better called Aether, the Akashic matrix or medium that underlies the physical manifestations of the world.

Here we elaborate the above concept in a rigorous quantitative fashion. We introduce the mathematics and the geometry of space-resonance and space-time, traditionally known as the Aether and identified in this book as the cosmic matrix, the Akasha. We show that the Aether/Akasha is a fabric of quantum rotating magnetic fields with electromagnetic, electrostatic, and gravitational dipole structures. The Aether Model identifies subatomic particles as angular momentum encapsulated in a quantum rotating magnetic field.

The APM is based on the same empirical data as conventional physics. However, the APM goes much further than conventional physics, pursuing the ultimate goal of unifying the four known universal fields or forces: the strong, the electromagnetic, the weak, and the gravitational forces. The APM unifies these forces through a set of general laws rooted in the fabric of space-time. The model accurately predicts the fundamental constants, the relative strengths of the forces, and the ground state electron binding energy for all the elements. All quantum, atomic, and molecular processes can be precisely modeled, leading to a new physics conveying fresh understanding and insight.

The Aether Model

The Aether Model refers to the same data as conventional physics' Standard Model; the data do not change, but the interpretation of the data does, as well as the tools for analyzing them.

In conventional physics one dimension of mass relates to one dimension of charge. In the APM, one dimension of mass relates to two dimensions of charge. Instead of expressing the unit of charge as Coulombs, it represents the unit of charge as Coulombs squared.

Conventional physics recognizes only one quantum of charge, the elementary charge. The elementary charge in the APM is noted as e^2, as opposed to e in the Standard Model. Furthermore, there is a second quantum charge, the quantum magnetic charge. There is a quantum magnetic charge for each type of particle which is directly proportional to the particle's mass.

The description of physical constants

The constants of conventional physics are, for the most part, not disputed by the Aether Model. Constants express values of which the dimensions have physical meaning. We therefore begin by introducing the description of physical reality based on the currently known physical constants.

By systematically analyzing the known universal constants, we can infer the *causes* of the constants from the constants discovered in the APM. We can describe the nature of these constants using dimensional analysis. We can further infer the causes of the new constants so as to discover what preceded them, and through dimensional analysis determine the nature and meaning of the new constants. We could then start from the beginning with the most fundamental constants and describe the course of evolution to the present condition of the universe.

We find that all constants, as pieces of a larger puzzle, originate from a Singularity represented by the dimensionless number 1. This is not the Singularity described by Big Bang theory, but more like that anticipated by ancient religions. The Singularity is a state where there are no dualities, no space, no time, and no matter. And yet, it is complete and whole.

The Singularity splits into two: (1) into the Gforce and resonating dark matter, and (2) into the magnetic and the electric charge.

$$1 = \frac{m_a \cdot \lambda_C \cdot F_q^2}{Gforce}$$

$$1 = \frac{e^2}{e_a^2 \cdot 8\pi a} \qquad (1)$$

(continues at table on next page)

m_a (Quantum Aether Mass – reciprocal to physical mass)	$m_a = 3.268 \times 10^{15} kg$	e (Elementary Charge)	$e = 1.602 \times 10^{-19} coul$
λ_C (Quantum Length, also Compton Wavelength	$\lambda_C = 2.426 \times 10^{-12} m$	e_a (Aether Magnetic Charge – reciprocal to e)	$e_a = 2.241 \times 10^{4} coul$
F_q (Quantum Frequency – reciprocal time)	$F_q = 1.236 \times 10^{20} \frac{1}{sec}$	a (Aether Fine Structure) [note: $a \neq \alpha$]	$a = 2.034 \times 10^{-48}$
Gforce (Reciprocal Force)	$1.21 \times 10^{44} newton$		

(1) The Gforce is equal to a reciprocal force:

$m_a \cdot \lambda_C$ is a string of mass

F_q^2 is frequency squared, or resonance

Dark matter consists of resonating strings of mass outside the fabric of space–time. These strings are like the seeds of physical existence, which when captured in the Aether unit, produce a stable subatomic particle.

(2) The Singularity also splits into magnetic charge (e_a^2) and electric charge (e^2). These charges are reciprocal to each other and have different geometries ($8\pi a$). [$a \neq \alpha$; the Aether's fine structure is not the same as the fine structure of electrons]. Furthermore, the magnetic charge is the charge carrier at different scales, manifesting as the strong nuclear force, permanent magnetism, and the Van der Waals force. The electric charge in turn is the carrier of the electric force. The different geometries of these charges separate these forces similar to a stick holding apart two halves of a clam shell. The magnetic and electric charges are orthogonal and tend to neutralize each other, giving rise to the weak

interaction (the weak interaction is a ratio of forces, not a force in itself).

$$A_u = Gforce \frac{\lambda_C^2}{e_a^2} \qquad (2)$$

When the Gforce acts on the magnetic charge, it produces a quantum unit of the Aether. This unit is a quantum rotating magnetic field. If the surrounding Aether density is not excessive, the Gforce maintains this rotating magnetic field indefinitely. When the Aether density exceeds the critical threshold, the Aether unit collapses like a burst bubble.

The Aether unit has five spatial–temporal dimensions consisting of three dimensions of length (space) and two dimensions of frequency (resonance). The aggregate of all Aether units likewise possesses this five–dimensional spatial–temporal structure.

$$A_u = k_C \cdot 16\pi^2$$

$$k_C = c \cdot Cd \frac{\mu_0}{\varepsilon_0} \qquad (3)$$

The Aether unit gives us Coulomb's constant. Coulomb's constant is composed of four physical constants: the speed of light, the conductance constant, the permeability constant, and the permittivity constant. The conductance constant provides one element of the explanation of the nonlocality discussed by Laszlo.

When the Aether unit absorbs a quantity of dark matter (string of mass), the combination yields a subatomic particle. Taking the case of the electron, which has angular momentum equal to Planck's constant (h), we obtain a structure similar to primordial dark matter:

$$h = m_e \cdot \lambda_C \cdot F_q^2 \qquad (4)$$

The electron itself is equal to, and it actually *is,* its angular momentum. Since the Aether unit encapsulates a string of mass, the Aether imparts to dark matter electrostatic and magnetic charges and a fixed geometry, thereby transforming the unit into visible matter. The same process gives rise to the proton.

The quantity of the mass in an electron and a proton are strictly determined by their position within the Aether unit. The quantum of strong charge of the electron and the proton are proportional to their masses as determined by the conductance of the Aether and the angular momentum of the particle.

$$e_{emax}{}^2 = Cd \cdot h$$

$$e_{pmax}{}^2 = Cd \cdot h_p \qquad (5)$$

The Gforce gives rise to the constants of force. When the Gforce acts on the strong charge of the Aether it gives rise to the magnetic constant, and this is the Aether unit itself. When the Aether unit is divided by its geometrical constant of $16\pi^2$, the Gforce relates to the electric constant (a.k.a. Coulomb's constant.) And when the Gforce acts on the mass associated with the Aether, it produces Newton's gravitational constant:

$$A_u = Gforce \frac{\lambda_C{}^2}{e_a{}^2}$$

$$k_C = Gforce \frac{\lambda_C{}^2}{16\pi^2 \cdot e_a{}^2} \qquad (6)$$

$$G = Gforce \frac{\lambda_C{}^2}{m_a{}^2}$$

The weak interaction is not an actual force; it is the ratio of the electrostatic force carrier to the magnetic force carrier. In the case of the electron, the weak interaction has a value equal to $8\pi\alpha$, where α is the electron's fine structure.

$$\frac{e^2}{e_{emax}{}^2} = 8\pi\alpha \qquad (7)$$

The Gforce is the common denominator of the three fundamental forces. The Aether constant is not only the magnetic force constant; it is also the quantum of space and possesses a specific double loxodrome geometry with the curvature constant of $16\pi^2$, which is equal to the spherical constant squared.

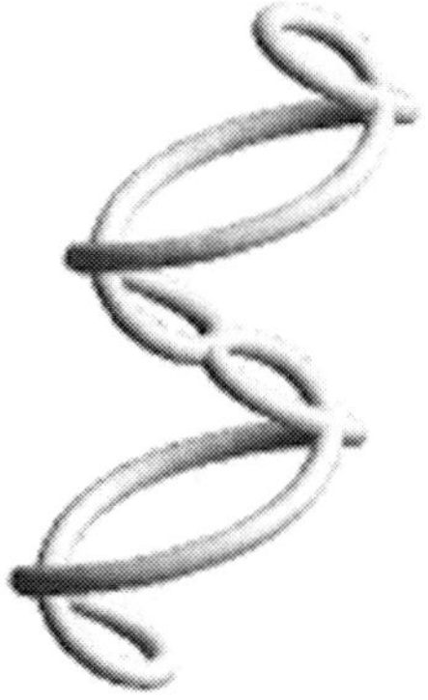

Double Loxodrome

The spring–like double loxodrome geometry of the Aether unit allows space to stretch, twist, and become deformed in numerous ways, which can explain many exotic quantum effects and macro effects, such as gravitational lensing. And because the Aether unit's geometrical constant is equal to the spherical constant (4π) squared, quantum particles and light can be quantified using spherical geometry, as demonstrated in the work of Milo Wolff.

Einstein's equation of General Relativity simplifies to

$$G = 8\pi T \qquad (8)$$

This basically states that the space–time curvature tensor (G) relates to the mass–energy tensor (T) by the geometric constant of 8π. This is equivalent to the electrostatic and magnetic charge relationship in the APM.

$$e^2 = 8\pi \cdot ({e_{emax}}^2 \cdot \alpha) \qquad (9)$$

The electrostatic charge is a property of the Aether, which is "donated" to matter. The magnetic charge is a property of matter. The electrostatic charge of space (space–time curvature tensor) relates to the magnetic charge of matter (mass–energy tensor) as adjusted by its fine structure. The 8π geometrical constant is due to 1–spin spherical–angle electrostatic charge acting on half–spin $\left(\frac{1}{2}\right)$ steradian–angle $\left(\frac{1}{4\pi}\right)$ magnetic charge.

The weak interaction is also shown to be a constant tug between the magnetic charge of matter and the electrostatic charge of the Aether. The weak interaction is therefore also a General Relativity effect.

Conclusions

The Aether Model, which we could only sketch here in its bare essentials, shows that the Aether (or ether) is the fundamental field or matrix for all the manifestations of the physical universe; it is the physics–formulation of the Akasha, originating in the perennial wisdom traditions. Thus a model of the physical universe based on the Aether is a new—or if you wish the rediscovered and reformulated timeless—paradigm in physics. It satisfies the fundamental ideal of science, namely, to produce the simplest and most consistent scheme for explaining our observations and tying together our findings.

References

David W. Thomson and James D. Bourassa, "Secrets of the Aether" (Aenor Trust, Alma, IL, 2004) www.secrets–of–the–aether.com

David W. Thomson and James D. Bourassa, "A New Foundations for Physics" (Paper submitted to Physical Interpretations of Relativity Theory conference, Imperial College, London, September 2006) http://www.16pi2.com/joomla/white–papers/doc_details/1–a–new–foundation–for–physics.html

David W. Thomson and James D. Bourassa, "Calculations of the Unified Force Theory" (Quantum Aether Dynamics Institute, Alma, IL, October 5, 2006) http://www.16pi2.com/joomla/white–papers/doc_details/3–calculations–of–the–unified–force–theory.html

Misner, Thorne, and Wheeler, *Gravitation*. 2003. W.H. Freeman & Company, 25th Printing, p. 380.

Summing Up*

We have noted that scientists have difficulty saying just *what* is the object of their investigation—what are quanta, what is life, what is mind, and what, in the final count, is the nature of the reality to which all these things refer. The shift to a new paradigm should facilitate finding credible answers. Since the paradigm suggested in this study claims that the world to which our theories refer is not just the manifest world, but includes a deeper, hidden but ultimately real domain, we should begin this "bottom line" summing up by saying just what is that domain.

What is the Akasha?

The Akasha is a dimension in the universe that subtends all the things that exist in it. It not only subtends all things: it generates and interconnects all things, and it conserves the information they generated. It is the womb of the world, the network of the world, and the memory of the world.

This concept is important for us, and it is important for science. It is important for us because it orients our thinking and our behavior. And it is important for science because it enables the integration of the various theories of the empirical disciplines into what Einstein called the simplest possible and yet comprehensive scheme—a scheme that gives us a meaningful picture of the world based on all of the accessible evidence we have of the world.

* Based in part on the Conversation with Györgyi Szabo published in the author's *The Birth of the Akasha Paradigm* akashaparadigm.com

How does this concept differ from our habitual picture of the world?

The Akasha is a very different concept from that which most people hold in the modern world. Understanding the world through the lens of the Akasha calls for a veritable "Gestalt switch." We normally think of the things we experience as real, and the space that embeds them as empty and passive, a mere abstraction. We need to turn this around. It is the space that embeds things that is real, and the things that disport themselves in space that are secondary. They are the manifestations of space, more exactly, of the Akasha, the underlying generative and interconnecting matrix that fills space.

This concept emerges from the findings of cutting–edge physics. Space, quantum physicists realize, is not empty and passive; it is a filled and active plenum, even though physicists still refer to it as the "*vacuum.*" In the emerging view space is the ground, and the things we know as real things in the world are the figures on the ground. They are figures not just *on* a ground; they are figures *of* the ground. The things we consider real are manifestations of space—manifestations of the cosmic matrix that fills space.

There is a good metaphor for this concept of the world. Think of waves traveling over the surface of the sea. When you look at the surface you see waves moving toward the shore, waves spreading out behind ships, waves interfering with waves. The waves move from one point on the sea toward another, yet there is nothing in the sea that would move that way: the molecules of water on the surface do not move from one place to another, they just move up and down. The motion of the waves is an illusion—an illusion not in the sense that there would be nothing that would correspond to it, but in that it is not what it appears to be. The waves travel across the surface of the sea, but the water of the sea does not travel.

The same applies to the motion of things in space. Things do not move *across* or *over* space, they move *in*, more exactly *within,* space. They are *conveyed* by space. Space is not empty and passive: it is the cosmic matrix that generates and connects all that exists. Everything that exists in the world exists within the dimension of the world we habitually call space.

The view this view gives us is very different from the commonsense view. The world that meets our eye is not an illusion, but it is not what it appears to be. The real world is not an arena of separate things moving across intervening space; it is a manifestation of a cosmic matrix. All things are part of that matrix, and are conveyed in and by the matrix. Not the bare existence of things is the illusion, but their separateness. All things are *in* and *of* the matrix, and in the final count are one with the matrix.

Can we be certain that this is the correct view of the world?

We cannot be *absolutely* certain, but we can be *reasonably certain.* There is no absolute certainty in science beyond the formulas of logic and mathematics. It is only there that we can have proof of the truth of our conclusions, because the proof, the same as our whole chain of reasoning, is "axiomatic": it is defined in its own terms without reference to anything else.

Einstein pointed out that as far as the proofs of mathematics do not refer to reality they are certain; and as far as they refer to reality they are not certain. Abstract schemes can be certain, but they become uncertain when they are applied to the real world. Sir Karl Popper summed up the implication for science. Scientific theories, he said, are improvable but not provable. Already two–and–a–half thousand years ago Plato warned us that our ideas about the world are at best a likely story. At this point in the evolution of our insight into the nature of things, the Akasha concept, I believe, is the likeliest story.

Is the Akasha the intelligence of the world?

The Akasha is a kind of intelligence. In the traditional context we would call it the spirit or consciousness of the world, and in the modern context we can consider it the logic or "program" of the world. It is what makes the world intelligible, makes the stars and planets, and the atoms and organisms behave in a way we can comprehend.

Is this a scientific view, or is it a spiritual view?

It is both. Genuine spirituality has always been based on the recognition of a deeper intelligence at work in the cosmos. The prophets and teachers of the world's religions gave their own interpretation of this intelligence, identifying it in ways that accorded with the concepts and symbolisms of their time. But a literal interpretation of their scriptures would be mistaken, for that would suggest that only their particular interpretation is true and valid. This would be similar to the dogmatic claim in science that only the data conveyed by the senses gives true information on the world; anything beyond that is idle speculation—"metaphysics." A mature science recognizes that the world is far greater and deeper than our sensory experience of it, just as a mature religion recognizes that the higher or deeper intelligence their doctrines suggest is the core of their system, and not the literal way their doctrines interpret it.

The insight conveyed by the Akasha paradigm is not in conflict with, and is indeed fundamentally the same, as the insight conveyed by the religious and spiritual traditions. A mature science is spiritual, and a mature religion is scientific. They are built on the same experience, and reach basically the same insights.

Can we experience the Akasha, the intelligence of the world?

The intelligence we identify with the Akasha in–forms all things in the manifest world. It in–forms our brain and body, and because our brain and body are our matter–like aspects, and we also have a mind–like aspect, Akashic in–formation appears in our consciousness. But it does not necessarily appear in our *waking* consciousness. If that in–formation is not to be repressed, or dismissed as fantasy and imagination, we need to enter what psychologists call an altered state of consciousness.

This was not always necessary. Shamans and medicine men and women, prophets and spiritual teachers have been able to access this cosmic intelligence as part of their everyday experience. They were frequently in altered states of consciousness: they cultivated them consciously, by rhythmic dancing, chanting, drumming, psychedelic herbs, rites and rituals, and disciplined meditation among other things. But in the modern world we have been so thoroughly convinced that everything we experience must come through our senses that we do not seek experiences that do not involve sensory apprehensions. And if such experiences occur spontaneously, we dismiss them as illusion or fantasy. The exceptions to this rule are artists, sensitives, spiritual people, and creative individuals in all walks of life, even in science. Often their key insights came in altered states, in prayer and in meditation, in esthetic experience, or in dreams, daydreams, hypnagogic states between sleep and wakefulness, and in deep introspection.

When we enter an altered state of consciousness elements flow into our consciousness that transcend the range of sensory perceptions. These elements are part of the totality of the information in the cosmic matrix we call the Akasha. This information is in a distributed form, as in a hologram. That means that all elements of the information are present in every part of it. In delving deeply into our consciousness, we access a holographic fragment of this information—of the information that "in–forms" the universe. In a modest but real sense, we "read the Akashic Record"— the record of all things that are, and have ever been, in the whole wide world.

The bottom line: What do we now know that we did not know before?

In the final count, science is about knowing: knowing the world and all things in the world. A science based on a new paradigm should convey more knowledge than a science based on the old. How does the Akasha Paradigm do in this regard? What knowledge does a science based on this paradigm give us about the world that a science based on the old paradigm does not and cannot give?

In his 2009 book devoted to querying the "true nature of the universe," biologist Robert Lanza of Wake Forest University created a kind of report card called "Classic Science's Answers to Basic Questions."* According to Lanza's report card, science is failing: it gives no knowledge at all in regard to eleven of thirteen basic questions, and it conveys a tentative answer to the twelfth and a negative answer to the thirteenth. Would a science based on the Akasha Paradigm do better? Here we contrast Lanza's questions and the answers he gives for "Classic Science" with the answers we can now offer for an Akasha-paradigm based science.

Q. How did the Big Bang happen?
A. *Classic science:* Unknown. *Akasha Paradigm science:* through the action of the repulsive forces that arose when the collapsing prior universe reached quantum dimensions.

Q. What was the Big Bang?
A. *Classic science:* Unknown. *Akasha Paradigm science:* a phase transition between universes, or phases of the metaverse; not a Big Bang, but a Big "Bounce."

Q. What, if anything, existed before the Big Bang?
A. *Classic science:* Unknown. *Akasha Paradigm science:* a preceding universe, or phase of the metaverse, with a physics similar to that of our own.

Q. What is the nature of dark energy, the dominant entity of the cosmos?
A. *Classic science:* Unknown. *Akasha Paradigm science:* not yet known, but offering orientation for the search: the answer is likely to be in the constitution of the Akasha, the physically real basis of spacetime and background of matter and energy in the universe.

*Robert Lanza, with Bob Berman. *Biocentrism: How Life and Consciousness are the Keys to Understanding the True Nature of the Universe.* Dallas, TX: BenBella Books, Inc., 2009.

Q. What is the nature of dark matter, the second most prevalent entity?

A. *Classic science:* Unknown. *Akasha Paradigm science:* the same as above—not yet known, but offering orientation for the search: the answer is likely to be in the constitution of the Akasha, the physically real basis of spacetime and background of matter and energy in the universe.

Q. How did life arise?

A. *Classic science:* Unknown. *Akasha Paradigm science:* the processes we see as basic to life originated as coherent relations emerged in time in the rich welter of organic molecules on the watery surface of some satellites in orbit around active stars.

Q. How did consciousness arise?

A. *Classic science:* Unknown. *Akasha Paradigm science:* Consciousness did not "arise"—it has always been present as an aspect of the intrinsically psychophysical universe.

Q. What is the nature of consciousness?

A. *Classic science:* Unknown. *Akasha Paradigm science:* it is the mental, more exactly "mind-like," aspect of the psychophysical universe, and of all the things that emerge and evolve in this universe.

Q. What is the fate of the universe; for example, will it keep expanding?

A. *Classic science:* Seemingly yes. *Akasha Paradigm science:* with a high probability it will reach a balance between the initial force of expansion and the gravitational force of contraction, after which it will re-contract to quantum dimension—to re-emerge as the next universe (or next phase of the metaverse).

Q. Why are the [physical] constants the way they are?

A. *Classic science:* Unknown. *Akasha Paradigm science:* the values of the constants have been progressively evolved in the preceding cycles of the metaverse and were passed on to our universe in the phase-transition we recognize as a Big Bounce.

Q. Why are there exactly four forces?

A. *Classic science:* Unknown. *Akasha Paradigm science:* The universal forces are not limited to four: the force (or field) responsible for nonlocal interaction is as basic and universal as the four classical fields; and there are various quantum fields and forces as well.

Q. Is life further experienced after one's body dies?

A. *Classic science:* Unknown. *Akasha Paradigm science:* after one's body dies not life as we know it seems to be experienced but a form of consciousness—encoded in ways yet to be discovered in the Akasha, which is the holographic in-formation matrix of the universe.

Q. Which book provides the best answers?

A. *Classic science:* There is no single book. *Akasha Paradigm science:* this answer is correct— there is no single book, nor is there ever likely to be one, although some books have always offered better answers than others and this is likely to be the same in the future.

Comments

Edgar Mitchell

In the very ancient past the natural human desire to understand our world, its content and the interactions in nature were stimulated by exploring beyond the local environment, discovering new flora and fauna and perhaps another tribe of humans living in a locale different from one's own. New words, new thoughts and new relationships presented themselves to mind and needed description and discussion. Experiencing nature and new beings expanded our understanding and allowed the creation of beliefs about how all life fits together in the larger scheme of things.

Looking back from the modern present we can chronicle the arising of many languages and many cultural beliefs in and about our world. The ancient mystics and wise men in every culture were the leaders in prescribing the nature of reality as they saw it, and the rules and procedures that should be followed in interacting with each other and with nature. Local cultural beliefs turned into local religions that claimed to speak the Truth about nature and the cosmos, including stories of the origins of all things and rules of behaviors humans needed to follow for successful lives and a successful social order.

As long as we remained local, such tribal procedures, beliefs and rules sufficed to regulate and inform our societies. The beginning of distant travel and interaction with remote peoples brought in the short run conflict and strife, more often than peace and harmony.

Enter the Middle Ages in the Western world where kingships, empires and travels to different continents became common. The Christian religion dominated Europe and set the standard for thinking and social interaction in the emerging nations. Disagreeing with the Church could be labeled as heresy and bring on death by burning at the stake. Then along came the philosopher, mathematician and thinker René Descartes. He wrote that body and mind, the physical and the spiritual belonged to different realms of reality that did not naturally interact. The Church authorities accepted this idea, which allowed Europe's intellectuals free thinking as long as they avoided the subject of mind and consciousness, the province of the theologians.

Shortly thereafter, Sir Isaac Newton published his laws of motion. Modern science and the classical laws of physics were born. Newton's physics was based on the conviction that experiment and testing were necessary to prove their application to reality. Interactions in nature had to be measured and proven with mathematical precision. This classical period in Newtonian physics lasted for four hundred years, until the turn of the 20th century, when Albert Einstein changed our understanding of the nature of space and time, and Max Planck, Erwin Schrödinger and Paul Dirac brought in quantum physics as a necessary element for our understanding of interactions at the subatomic level.

The particular rules and the basic beliefs people had used to understand the nature of reality constitute what philosopher Thomas Kuhn called a *paradigm*. Descartes, followed by Newton and his laws of physics, initiated a 400+ year reign in which classical physics was the paradigm for science. The discovery of the quantum world and the codification of its interactions in the 1920s ushered in changes that launched a paradigm–shift that took up most of the twentieth century. However, the two revolutionary theories, general relativity and nonlocal quantum mechanics, have not been successfully unified and the post–Newtonian paradigm lacked integral consistency. It has become clear, on the other hand, that the Cartesian separation of matter and mind is a faulty concept: both matter and are basic elements of reality. Research on consciousness surfaced by the end of the 20th century as an important topic in science after four centuries of neglect.

Today, the understanding of reality wrought by the rise of quantum theory, and the accumulation of evidence regarding the active role of consciousness in the world, catalyze a basic paradigm–shift. This is what Laszlo's new book is all about. What it accomplishes merits our serious and urgent attention.

Bringing to the surface science's dominant paradigm and examining its power—or its failure—to provide an integral and realistic understanding of the world is one of the basic accomplishments of Laszlo's new book. The second is to outline the new paradigm that could overcome the shortcomings of the old and endow the theories of the contemporary sciences with integral meaning and realism. Science has outgrown its dominant paradigm. We need a new paradigm to understand the world that emerges at the cutting–edge of the sciences. Laszlo's Akasha paradigm fulfills this requirement. This is a revolutionary accomplishment of enormous scope and indisputable relevance. We need to take account of it, discuss it, and follow it up with sustained research.

David Loye

Behind Ervin Laszlo's brilliant development of the Akashic Paradigm lies the long history of efforts to understand who we are, where we've been, and what happens next for us in evolution.

Among the ancient Hindus, Mayans, and over and over again in history what emerges is the picture of the clash of the vast, over–riding mindsets we call paradigms. For a long time one paradigm will rigidly hold most of us captive, then out of a time of cataclysmic trouble and confusion—such as we're experiencing now—a new paradigm will emerge.

Where Laszlo's Akasha paradigm fits into this picture is with the great shift in our time toward a bold new perspective that embraces, then weds and transcends both the old paradigms of religion and the paradigms of mainstream modern science.

With the Akasha paradigm we are entering what might be called an Age of Reconciliation. Rooted in a vast understanding of the heady dynamics of advanced physics, the Akasha paradigm weds the best of religion and the best of science into a powerful expression of a new partnership between progressive science and progressive religion to beat back the forces that are trying to drive us backward in evolution. Laszlo's grounding for the kind of mindset we need to gain a better future has been examined and blessed by more leading scientists than any other I know of.

Kingsley Dennis

At each stage of human evolution we are confronted by phenomena that call for us to investigate our paradigms of knowledge and understanding. At each stage we are asked to rise to the challenge to conceptualize, visualize, and vocalize these insights that are necessary to compel us forward along the unfolding of social evolution. How we navigate our present and potential future(s) is fundamental to how we survive as a species co–dwelling on a vibrant and life–sustaining planet. It is clear that our incumbent frames of scientific knowledge are in need of timely re–adjustment. Ervin Laszlo's presentation of the Akasha paradigm does exactly that.

The Akasha paradigm returns our way of thinking to an integral consciousness; a non–linear mode of understanding that prompts us to accept the reality of nonlocal interactions. This view shifts us away from the dominant mental–rational worldview toward a perspective that fosters an ecological–reciprocal relationship as we exist within an inclusive whole. This calls for nothing less than a mutation in human consciousness.

By adopting the Akasha paradigm we are bringing greater meaning into our life. This paradigm does not destroy, or collapse, our current or older models; rather, it updates our previous stages of knowledge into a more inclusive model that better serves to explain how the manifest world exists—and *can* exist—within a 'hidden dimension' that underlies the structure of a more complete, inclusive energetic reality. The Akasha paradigm allows for the existence of materiality whilst simultaneously embracing the unification of all known phenomena—it is a unity paradigm at heart; and as such it is life–sustaining in its implications for humankind.

The Akasha paradigm as set forth by Ervin Laszlo tells us that life exists within a pattern that is coherent beyond our wildest dreams. Living systems seek harmony throughout their evolutionary journey. This insight uplifts our spirit and imbues our life with new meaning.

Stanley Krippner

The standard cosmology of the 20th century told a story starting with the Big Bang, a single event that created the cosmos. Eons later, gigantic swirls developed that became galaxies. Eventually, at least one solar system emerged that contained a planet capable of developing and sustaining life.

This story supplanted earlier stories that attributed creation to deities (or a single deity), and invented ingenious mechanisms such as epicycles to preserve the notion that Earth was the center of its solar system. When this story unraveled, it was not an easy matter to tell another one. One early story–teller paid with his life while another was put under permanent house arrest.

The 20th century story tellers look back at those stories as superstitious tales, taking pride in the ascent of science as a guide to constructing creation tales that were based on reasoned logic, empirical observation, and carefully constructed experiments. However, it was these very tools that revealed that the 20th century story had its own epicycles, mechanisms that did not hold up once quantum mechanics entered the picture.

In this brilliant book, Ervin Laszlo tells a story for the 21st century, a narrative inspired by a much earlier story from ancient India, the Akashic Records, perhaps the first "theory of everything" that highlighted nonlocality, a term absent from the 20th century standard cosmology but paramount a century later. When physicists observed that separated subatomic particles at the microlevel continued to interact at impossible distances, this phenomenon was dismissed as an interested but unimportant quirk. In any event, it could not be repeated at the macrolevel.

However, the data began to accumulate from physics, chemistry, biology, and the disreputable field of parapsychology that nonlocal events spanned the gap between micro and macro. Time, space, and the ephemeral construct called "consciousness" engaged in a dance that quantum physicist David Bohm called the "holomovement." Indeed, Bohm was one of the first to provide a narrative that was dismissed at the time but that probably insured his place in the history of the philosophy of science.

Drawing upon Bohm, Schrödinger, Man–Wan Ho, and many others, Laszlo's story tells us that the Akashic world is a locally as well as a nonlocally interconnected and interacting world. The Akashic world includes a dimension in the universe that subtends all the things that exist in it. It not only subtends all things, it generates and interconnects all things.

Some people of a particular doctrinaire bias will read this story and will assume that this dimension supports such fairy tales as Creation Science or Intelligent Design. Far from it. The Akashic cosmos is a self–organizing whole that creates itself. It is reminiscent of Charles Darwin's closing statement in *The Origin of Species.* Darwin ended his story by commenting, "It is interesting to contemplate a tangled bank, clothed with many plants of many kinds, with birds singing on the bushes, with various insects flitting about, and with worms, crawling through the damp earth, and to reflect that these elaborately constructed forms, so different from each other, and dependent upon each other in so complex a manner, have all been produced by laws acting around us."

This is as close as Darwin could come, given his era and his knowledge, to describing nonlocality. But he was on target when he exclaimed, "There is grandeur in this view of life." And in his remarkable update of Bruno, Copernicus, Galileo, Darwin, Einstein, and others, Ervin Laszlo, heeding Plato's warning that even the best designed account of the nature of reality is but a likely story, has now good reason to tell us, *the Akasha paradigm is the likeliest story we can tell today.*

Deepak Chopra

The Akasha is not the electromagnetic field or a physical field but literally a realm of transcendental consciousness, the consciousness from which the whole universe arises and into which the universe again subsides.

Understanding Laszlo's work on the Akasha paradigm gives insight into platonic values like truth, goodness, beauty, harmony, and ambition. These give rise to spontaneous and not imposed morality, coming from an experience of our higher self. They give scientific insight into the most fundamental nature of the universe. At the fundamental level the universe is more than space, time, and energy, spin and charge, and all the things physicists talk about. The Akasha is the fundamental building block of an evolving, maturing mind. Understanding it opens the way to a higher morality and a higher consciousness.

CPSIA information can be obtained at www.ICGtesting.com
Printed in the USA
LVOW110920220512
282766LV00001B/3/P

9 781937 504182